GEODESIGN, URBAN DIGITAL TWINS, AND FUTURES

Geodesign, Urban Digital Twins, and Futures explores systems, processes, and novel technologies for planning, mapping, and designing our built environment. In a period of advancing urban infrastructure, technological autonomy in cities, and high-performance geographic systems, new capabilities, novel techniques, and streamlined procedures have emerged concurrently with climatic challenges, pandemics, and increasing global urbanisation. Chapters cover a range of topics such as urban digital twins, GeoBIM, geodesign and collaborative tools, immersive environments, gamification, and future methods. This book features over 100 international projects and workflows, five detailed case studies, and a companion website. In addition, this book examines geodesign as an agent for collaboration alongside futuring methods for imagining and understanding our future world.

The companion website for this book can be accessed at http://geodesigndigitaltwins.com.

Paul Cureton is Director of The School of Design, and Senior Lecturer in Design at ImaginationLancaster, and a member of the Data Science Institute (DSI), Lancaster University.

Elliot Hartley is a 3D GIS, digital twin and development planning professional and internationally recognised 3D geodesign expert.

"Ever since computers were invented, designers have been energised to use them to create more liveable and sustainable cities. But only recently have new methods emerged to help us think about this future. This book introduces those at the cutting edge: new visualisations, information management, digital twins, and geodesign, all mediated in environments where participation is central and essential. This is a book for all those who believe that contemporary computing is essential to the future of urban planning, and that must be all of us."

Michael Batty, *Centre for Advanced Spatial Analysis, University College London*

"This book provides a rich academic resource for understanding digital twins and geodesign and the amazing benefits they bring when used together. Many cities are already leveraging living urban digital twins to improve their day-to-day operations. Digital twin implementations are just beginning and are increasingly being used in cities of all sizes. It is my belief that the integration of geodesign is increasingly necessary, using the power of geography to integrate information, and allowing planners, designers, and citizens to participate. This book provides powerful examples and references for how digital twins, along with geodesign, help us to understand, model, and design a better future."

Jack Dangermond, *President and Founder, Esri*

GEODESIGN, URBAN DIGITAL TWINS, AND FUTURES

PAUL CURETON AND ELLIOT HARTLEY

Routledge
Taylor & Francis Group
NEW YORK AND LONDON

Designed cover: Elliot Hartley/GD3D®
Contains OS data © Crown Copyright and database rights 2024
Contains data from OS Zoomstack
Contains GD3D® Copyright 2024
Contains LCIM 3D Open Data

First published 2025
by Routledge
605 Third Avenue, New York, NY 10158

and by Routledge
4 Park Square, Milton Park, Abingdon, Oxon, OX14 4RN

Routledge is an imprint of the Taylor & Francis Group, an informa business

© 2025 Paul Cureton and Elliot Hartley

The right of Paul Cureton and Elliot Hartley to be identified as authors of this work has been asserted in accordance with sections 77 and 78 of the Copyright, Designs and Patents Act 1988.

All rights reserved. No part of this book may be reprinted or reproduced or utilised in any form or by any electronic, mechanical, or other means, now known or hereafter invented, including photocopying and recording, or in any information storage or retrieval system, without permission in writing from the publishers.

Trademark notice: Product or corporate names may be trademarks or registered trademarks, and are used only for identification and explanation without intent to infringe.

ISBN: 978-1-032-74862-7 (hbk)
ISBN: 978-1-032-74861-0 (pbk)
ISBN: 978-1-003-47131-8 (ebk)

DOI: 10.4324/9781003471318

Typeset in Akzidenz Grotesk and Franklin Gothic
by codeMantra

*For Tamara, Alana, Oscar, and Talia
Love, Play, Purpose, and Family.*

PP, Bro'ski, Lllillluuuppp, Besties#

Paul

For Michelle, Elizabeth, Katherine, and Juliette

Elliot

CONTENTS

P XII **ABOUT THE AUTHORS**

P XVI **ACKNOWLEDGEMENTS**

P 2 **INTRODUCTION**
GEODESIGN, URBAN DIGITAL TWINS, AND FUTURES

Geodesign, Urban Digital Twins, and Futures	p 3
Book and Tutorial Structure	p 7
Introduction to the Tutorials	p 8
Start Simple Ask the Right Questions	p 8
Be Realistic	p 9
What You Need: The Right Kind of Data	p 9
Tutorial Structure	p 10
Futuring	p 12
Geodesign	p 17
Data Acquisition, Access to Data, and Open Data	p 21
Smart Cities and Digital Twins	p 24
Gamification	p 30
Summary	p 32

P 42 **CHAPTER 1**
DEFINING THE LIVING LAB

Introduction	p 43
We Have the Technology, But What Is the Diagnosis?	p 53
TRL of Digital Twins and GIS – The Maturity Space	p 58
Models and Models and Models...	p 63
Urban Digital Twins and Embracing Speculation	p 67
World-Building	p 71
Limitations	p 77
Towards a Citizen Science – We Can Only Do This Together	p 78

P 90 **CHAPTER 2**
TOWARDS URBAN DIGITAL TWINS

Urban Digital Twins Overview	p 91
Towards Virtualisation of Complexity – Baseline, Frequency, and Data Collection	p 98
Urban Analytics	p 112
Dashboards	p 119
Digital Twin Interactions	p 122

P 136 **CHAPTER 3**
GEODESIGN AND URBAN DIGITAL TWINS

Introduction	p 137
History of Computer Graphics and Geodesign	p 145
Geodesign as a Framework for Working with UDTs	p 148
Assessment	p 151
Intervention	p 152
People	p 153
Models	p 153
Platforms	p 153
Summary	p 160

P 172 **CHAPTER 4**
GEODESIGN METHODS FOR URBAN DIGITAL TWINS

Introduction	p 173
Geodesign Methods – People, Models, Platforms	p 177
People	p 178
Backcasting – Geodesign	
Stage – Representation Model, Process Model	p 178

ix

    Data Stories – Geodesign
        Stage – Representation Model,
        Impact Model, Decision Model   p 181
    VGI – Geodesign Stage – Representation
        Model, Process Model   p 182
    Participatory GIS – Geodesign
        Stage – Representation Model,
        Process Model, Decision Model   p 184
Models   p 188
    Perception Studies – Geodesign
        Stage – Representation Model, Decision
        Model   p 188
    Procedural – Geodesign Stage – Evaluation
        Model, Change, Model, Impact Model   p 190
    Land Surface Model – Geodesign
        Stage – Process Model, Evaluation
        Model, Change Model   p 193
    Agent-Based-Modelling – Geodesign
        Stage – Change Model, Impact Model,
        Decision Model   p 195
Platforms   p 196
    Data Dashboards – Geodesign
        Stage – Change Model, Impact Model,
        Decision Model   p 196
    Augmented – Geodesign Stage – Process
        Model, Evaluation Model   p 199
    Physical Model – Geodesign
        Stage – Representation Model, Impact
        Model, Decision Model   p 204
Summary   p 208

P 216 **CHAPTER 5**

GAMING, WORLD-BUILDING, AND PARTICIPATORY PLANNING

Introduction	p 217
World-Building Games and Simulation Tools	p 218
Gaming Planning Systems	p 229
Urban Game Continuum	p 231
Modifying Cities: Skylines with GeoData, Lancaster City, UK	p 238
Limitations	p 244
Conclusion	p 248

P 260 **CONCLUSION**

GEODESIGN, URBAN DIGITAL TWINS, AND FUTURES AT THE EDGE

Conclusion	p 261
Visioning and Experimentation	p 273

P 282 **INDEX**

ABOUT THE AUTHORS

Dr. Paul Cureton, FRSA, is Director of The School of Design and Senior Lecturer in Design at ImaginationLancaster, and a member of the Data Science Institute (DSI), Lancaster University. His work transcends subjects in spatial planning, 3D GIS modelling, and design futures. It is at the forefront of exploring the critical interface of new and emerging socio-technological relationships such as *Design* for Digital Twins, Drone Futures, and novel process-based methodologies for Future Environments such as geodesign and XR interactions. His recent publications include the monographs, *Strategies for Landscape Representation: Digital and Analogue Techniques* (Routledge, 2016) and *Drone Futures: UAS for Landscape & Urban Design* (Routledge, 2020). He is also co-author with Nick Dunn of *Future Cities: A Visual Guide* (Bloomsbury, 2020).

Elliot Hartley is a 3D GIS, digital twin and development planning professional. He is also an internationally recognised 3D geodesign expert. He is also a pioneer and an instructor in the use of Esri's ArcGIS CityEngine for 3D city modelling and urban planning projects.

Elliot has accumulated more than 20 years of professional experience in the fields of GIS, planning, 3D urban modelling, and geodesign. He has served as a senior planner in private practice, as well as a Development Control officer within UK local authority planning departments.

He has consulted for and trained teams in some of the world's leading planning and architectural consultancies in Europe, North America, and the Middle East. In addition, he has also provided training and services for local government/city authorities in Europe, North America, and the Middle East.

As a subject matter expert in 3D urban modelling, Elliot has spoken and conducted workshops/webinars/seminars at numerous international events on topics of planning, 3D urban modelling, and geodesign.

Elliot has a Master's in Geographical Information for Development (MSc Durham) and Town and Country Planning (MA UWE), as well as a degree in Geography (BA Hons RHUL). He is also the Chair of Governors for a local Primary School. He also maintains a long running 3D geospatial and planning blog called GeoPlanIT.

ACKNOWLEDGEMENTS

This book is the result of an initial professional connection and collaboration, which also turned into a friendship and continuing dialogue. First driven by a shared interest in 3D GIS, procedural modelling via ArcGIS CityEngine, and reality capture from smartphones and drones, but also in terms of wider dialogues around urban digital twin (UDT) ambitions, the planning profession, and discussion of geodesign. In terms of the division of work, Paul has written this book and Elliot has produced the valuable learning resources.

Perspectives have been rich from Elliot's commercial practice and Paul's academic work, both seeking pragmatic, achievable routes for implementing UDTs. There is much to be done, huge barriers, silos, up-skilling, and resourcing for UDTs, yet there is a shared belief that UDTs can move beyond 'smart cities' and have a useful, tangible public good in the climate emergency. This project has been a shared motivation and has been reinforced through several achievements, such as establishing the Lancaster City Information Model (LCIM) model or witnessing children's sheer joy in workshops delivered and the creation of their own city models for participatory future cities.

Many of the ideas of this book for Paul stem from professional networks and dialogues sustained during the period of book production. In particular, the CIM Forum (City Information Modelling, September 12, 2023) and the contributors across the geospatial, planning, and design professions. Paul would particularly like to thank Lou Welham, Ministry of Housing, Communities and Local Government; Ralph Coleman, Bluesky International, Ordnance Survey; Gordon Blair, UKCEH; and Ellie Brown, Lancaster City Council and the GIS and Planning Teams, amongst many others, to which I am grateful for the connection. I am also grateful to two key influential publications from Li Wan, Timea Nochta, Junqing Tang, and Jennifer Schooling,

Digital Twins for Smart Cities: Conceptualisation, Challenges and Practices (2024) and Michael Batty, *The Computable City Histories, Technologies, Stories, Predictions* (2023).

Paul would like to thank Paul Coulton, Imagination Lancaster, for collaborating on the game design process, design fiction, and gamification, which has been incredibly fruitful in my research direction. Paul would also like to thank Nick Dunn, Imagination Lancaster, for continuing dialogues on the nature of futures and urban environments. Paul wishes to give special mention to Anna Jackman, University of Reading, for the thoughtful and constructive review of a draft chapter. Imagination Lancaster is an open exploratory design research lab, and this environment has enabled a multidisciplinary approach and research space for Paul to explore the topic of urban digital twins. Paul finally wishes to thank Vanessa Longden for their administrative support in this project and key research office support from Zoe Bolton.

This book was based on a number of prior publications. Paul would like to thank journal editors and Connor Clive Graham and Ahn Chaewon, NUS conference organisers. Four particular outputs have informed the creation of the book chapters listed below. Funding support stems from E3: Expanding Excellence in England: Beyond Imagination Project, ESRC and Digital Planning.GOV, MHCLG.

Chapter 1.

Cureton, P & Hartley, E 2023, City Information Models (CIMs) as Precursors for Urban Digital Twins (UDTs): A Case Study of Lancaster, *Frontiers in Built Environment*, vol. 9, 1048510. https://doi.org/10.3389/fbuil.2023.1048510.

Cureton, P, *World Building Urban Digital Twins: Designing for the Future Urban Model Making: Knowledge, Technology and*

Society, Asia Research Institute, National University of Singapore (NUS), 2nd–3rd, March 2023. https://ari.nus.edu.sg/events/urban-model-making/.

Chapter 3.

Cureton, P 2024, Geodesign for Environmental Resilience. In R Brears (ed.), *The Palgrave Encyclopedia of Sustainable Resources and Ecosystem Resilience*. Palgrave Macmillan, Cham. https://doi.org/10.1007/978-3-030-67776-3_34-1.

Chapter 5.

Cureton, P & Coulton, P 2024, Game-Based Worldbuilding: Planning, Models, Simulations and Digital Twins, *Acta Ludologica*, vol. 7(1). 10.34135/actaludologica.2024-7-1.18-36.

Elliot is enormously grateful for Paul's efforts in putting their extensive conversations on 3D urban modelling and geodesign into a cohesive practical text. Distilling the large amount of academic literature and bringing it back to practical applications is no easy task! Inspiration for being involved in this book and working with Paul came from client interactions whilst working at Garsdale Design/GD3D to which he is externally grateful. In addition, Elliot would like to thank his family for the support that they give.

Finally, but most importantly, Paul would like to thank his family for their extended support in writing this book, especially Gill Clark. Family has supported and encouraged writing and regulated caffeine intake, and we have shared beautiful family moments together during downtime. Paul has enjoyed all the achievements of the children, small and large, and it has been a particularly special time during this book writing.

INTRODUCTION
Geodesign, Urban Digital Twins, and Futures

GEODESIGN, URBAN DIGITAL TWINS, AND FUTURES

Over the last 30 years, a paradigm shift has occurred in terms of geography, computing, and design. This book is about the divergence and convergence of people in terms of how we can work together in terms of planning and adapting to climate change, models, and selecting the range of models that can help create predictions and platforms in terms of technologies that coalesce people and models around a common base for urban spaces. Not only has computing transformed how people work, but it has also delivered transformational shifts to disciplines and opened new avenues for collaboration. This period is the age of large volumes of ubiquitous data, in near-real-time, captured from embedded, connected, and remote sensors with the ability of participants to engage in various forms of immersion and interaction. Advancing urban infrastructure, technological autonomy in cities, and high-performance geographic systems, new capabilities, novel techniques, and streamlined procedures have emerged concurrently with climatic challenges, pandemics, and increasing global urbanisation. The book charts these paradigm shifts, advances, methods, and responses to these challenges.

During the 1970s, the Harvard Laboratory experimented with cartographic research problems and the use of computing for spatial analysis. Geoffrey Dutton created a hologram of the United States' population growth between 1790 and 1970. As John Hessler states, this cartographic research developed from early work by William Warntz on mathematical representations of topographical surfaces, ultimately fundamentally changing what it means to make a map (Hessler, 2009, p. 2) (Figure 0.1). What we 'select' regarding maps, statistical data, and cartography refers back to our 'twinning' of geographic realities. Dutton and others were part of a period of innovation and at the birth of modern computing in developing frameworks and experimenting with novel visualisation techniques such as

GEODESIGN, URBAN DIGITAL TWINS AND FUTURES

Figure 0.1
Geoffrey Dutton, The holographic Map of the United States at the Harvard Lab for Computer Graphics, 1978. An early example of four-dimensional dynamic cartography with the hologram rotating in space and changing with time. Photograph by John Hessler. Library of Congress.

Figure 0.2
Joy Division, Unknown Pleasures Album, 1979 by Factory Records. Kay Roxby/Alamy Photo. Designer Peter Saville created the album cover from stacked radio emissions data from the first pulsar in 1967.

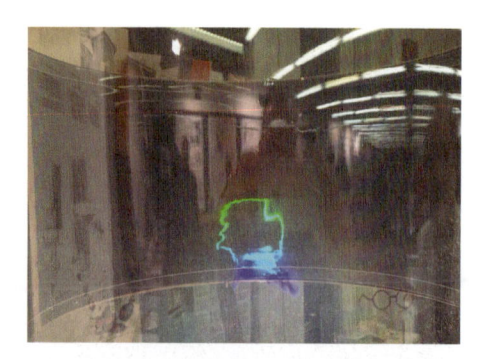

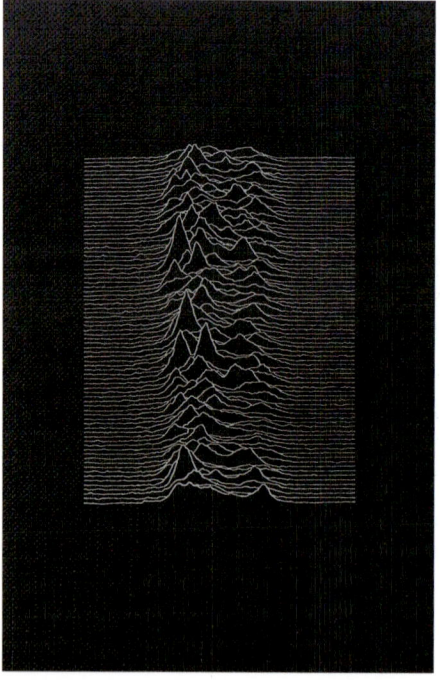

4D (time)-based holographic mapping, amongst other things such as pulsar radio emissions from neutron stars, famously appearing on the band Joy Division, album cover (Chrisman 2006) (Figure 0.2). Fast forward to today, and the European Space Agency (ESA), at the time of writing this book, is working on a 'Digital Twin' of Earth, a digital replica of our planet that is highly dynamic, concentrating on satellite observations, simulations, and artificial intelligence (AI) in the fields of forestry,

Figure 0.3
European Space Agency (ESA), Digital Twin of Earth, 2023, ECMWF/DestinE. The ESA is developing a dynamic virtual replica of the Earth based on Earth observation data, in-situ measurements, and AI.

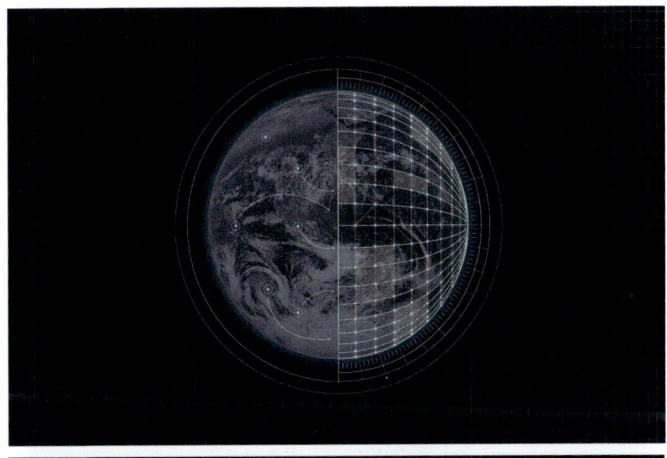

hydrology, Antarctica, food systems, and oceans and vital environmental data (Figure 0.3). Digital Twin Earth will utilise AI and embedded sensors to help visualise and forecast the impact of human activity. Between the examples of Dutton and the ESA, modern remote sensing and geographic information systems have developed substantially since the late 1960s and the range of innovation and experimentation should be accounted for. The capabilities of the digital twin of Earth can be juxtaposed with UN Habitat's findings that two-thirds of the world will be living in urban areas by 2050, meaning the role of geographic information systems (GIS) to highlight and fight climate change, spatial inequality and prosperity, and urban planning is incredibly important (UN Habitat, 2022). We can now monitor and model humanity's impact on the world around us in increasing detail. GIS for urban planning are essential for people-centred tools, and the UN-Habitat Participatory Incremental Urban Planning (PIUP) toolbox is one example of working with stakeholders to understand urban planning processes. PIUP is a step-by-step methodology that improves real-world cases at the time of writing, informing 17 projects across Africa and the global south. Ultimately, this book examines contemporary technology and its interfaces with society, which is termed social-technological, to transform urban environments sustainably with planning foresight to deliver efficient structures, processes, research, and implementation.

In urban planning, urban design, and landscape planning, there are several relevant areas, including data acquisition, use of GIS, cartographic mapping practices, analytics and visualisation, development workflows, design options, and scenarios, all part of a broader shift of digital transformation. By their nature, these areas have emerged from alternative disciplines, transdisciplinary approaches, or stem directly from GIS science. As Peter Hall states, spatial planning is intertwined with spatial representation, be it map or visualisation (Hall & Tewdwr-Jones, 2020, p. 4) as loose principles or for detailed works for physical construction. As such, the role of GIS and spatial databases is critical. Indeed, according to Hall, the planning activity with the onset of computer science may automate processes but does not alleviate planners' social and professional responsibility (Hall & Tewdwr-Jones, 2020, p. 8). Thus, many discussions around AI, automation, and digital systems need to include human relationships, forms of interaction, and ethics. These areas of enquiry cover areas of convergence central to this book's aim. These areas are geodesign, 3D geo-data and urban analytics, urban digital twins (UDTs), and gamification and bridge the socio-technical aspects. These convergent areas overlap and are porous, forming entire systems or are areas with a specialist enquiry that have only been partially or theoretically connected to date. The authors do not prescribe these areas as a necessity for delivering future urban planning and urban design. Indeed, place context and resourcing are critically important, and often, no premeditated goals or planning objectives have been set. However, the authors map this book's contemporary practices, future visions, and workflows so that readers can adapt, diversify, and implement their work.

BOOK AND TUTORIAL STRUCTURE

The authors have produced this work from previous collaborative projects and a shared interest in 3D GIS, modelling, and simulation from planning and other perspectives. *Geodesign, Urban Digital Twins, and Futures* appeals to multidisciplinary subjects, including architecture, landscape architecture, urban design, and urban and regional planning, utilising futures methods and GIS in the built environment. Such coverage is due to the convergence rationale that many traditional disciplines are moving to a post-disciplinary mode of working, sharing many techniques and approaches with ever-increasing knowledge overlaps. Many of these post-disciplinary approaches are due to the changing digital technology, AI, accessibility, interoperability, and a broader need to address the complexity and challenges of the built environment. For example, in the regional planning field, there have been various calls for critical readjustment of the profession. John Harrison et al. argue for regional planning futures, which need to be,

> *captured systematically and presented visually and accessibly to wider audiences. The data and digital tools increasingly available need to be harnessed as a continuous and enduring set of accessible and interpreted spatial analysis, not only led, and controlled by technologists, but used by spatial strategists and social scientists and available to all.*
> (Harrison et al., 2020, p. 7)

The authors identified a need for a comprehensive book covering these paradigm changes but also in an accessible way with real-world cases and tools for implementation in readers' projects to map their own future directions. For example, in an exploration of an open-source planning system for Stockholm County, Sweden, by Jessica Page et al. (2020, pp. 1509–1510), 'What If' future scenario methods were utilised but on a system that attempted to address three planning impediments of functionality, capacity to adapt, and governance. This book is, therefore, a 'playbook' of tactics for complex environments but also a book for strategic arrangements to overcome impediments. This book, therefore, aims to account for the latest developments in GIS science, urban technology, and urban futures. However, such an ambition for tutorials and instruction has a limited shelf life in book form due to the pace of change in digital technology; therefore, it has a companion website for news, techniques, and articles as a supplement to the book.

Introduction to the Tutorials

This book discusses many academic concepts of geodesign, UDTs, and urban futures, whilst highlighting practical applications and case studies. We do acknowledge that sometimes, if you are anything like us, you feel that the academic text makes you a little blurry! Perhaps it is the rich and deep referencing and citation of sources, or perhaps you are reading this because you want to be creative and do something. Some readers may ask: "What if I want to make my own 'Digital Twin'? What if I want to work on my own 'geodesign' project?". The desire to stop reading and just get on with it can be strong, so we have linked this to various tutorials.

Start Simple Ask the Right Questions

First, you need to figure out what it's all for. What question or issue are you trying to address? Is it designing a new urban area or monitoring environments? Also, ask yourself, who is this for? You, the reader, may have one idea of what a 3D model looks like in your head, we the authors guess at this point it's either a video game or Google Earth visualisation, yet from a professional perspective, working over 20 years with 3D tools, maybe you have quite a different idea of what a 3D model may look like. So, asking the right questions early on here is crucial because it helps you form a workflow or pathway dictating effort, time, and, of course, cost, and it is one of the first steps of geodesign.

Be Realistic

Three-dimensional technologies are exciting and engaging, but be realistic; sometimes, a 3D model is not the answer; sometimes, the answer is not even a map. While we may not like it visually, sometimes, the answer lies in an Excel sheet. Inevitably, for many who are working in professional contexts, the question isn't what do you think is best? It can often be dictated by a stakeholder in private practice, which may mean the client, but frequently, it's the public.

WHAT YOU NEED: THE RIGHT KIND OF DATA

It is the experience of this author that data that is free is often far from perfect, but equally, data that is paid for can be a disappointment, especially if you've paid a lot for it. Our experience of data, though, is rather interlinked with expectations, so before deciding on datasets, cast your mind back to those questions you want to answer. It is important to think of outcomes here. Here is an anonymised but fairly typical example of professional project work this co-author has worked on:

A client wants 'master plan' of specific urban area within an existing urban core of city. You have done the work in 2D (maps and CAD perhaps) as well as having crunched some numbers, and now you need a 3D model for visualisation to show how it will look for stakeholders (not just the client here but perhaps the public or local councillors).

The temptation here is to obtain beautiful, accurate 3D models of the existing city to provide context for our wonderful, innovative design. So, you do that but don't fully understand how these models are made. You ask for real-world textured imagery on those buildings because you know Google Earth has it… Now, you have the challenge of combining two datasets (potentially made for different end uses and/or software platforms).

If you're successful, you create a nice 3D visualisation of the new and old area. Except when you proudly display this 3D model, a nearby resident/local councillor/client notices that your expensive 3D real-world textured buildings are out of date or wrong. Now, the whole model is called into question, including your design! If you can't correct that existing building, what else did you get wrong?! A potential hypothetical project disaster, and one that we can assure you, has its origin in a very real-world project.

Now imagine the same scenario, except you sourced free open data of building footprints and extruded them (to be box-like) using height data from open lidar data. Same designed model as previously, but your contextual buildings are boxy extrusions and white. Now, stakeholders view this visualisation differently! Clearly not real, clearly not accurate and more diagrammatic! Would anyone notice? Would anyone care? No, probably not. Are we saying always go for the cheapest when sourcing data for your project? No. Are we saying you should understand the potential consequences of your choices beforehand? Yes.

Tutorial Structure

Fundamentally, you need data, then probably some software/hardware tools, and then an idea of how to combine them to create something useful. These are what our associated practical guides and tutorials are for. Each tutorial has been paired with a chapter in our book, linking to a theme or themes within each chapter.

- Chapter 1 – Defining the Living Lab
- Tutorial: 3D Modelling and Scanning
- Chapter 2 – Towards Urban Digital Twins
- Tutorial: The Making of a Lancaster CIM
- Chapter 3 – Geodesign and Urban Digital Twins
- Tutorial: Procedural Geodesign with CityEngine
- Chapter 4 – Geodesign Methods for Urban Digital Twins
- Tutorial: Data Walks, one of eleven Geodesign methods
- Chapter 5 – Gaming, World-building, and Participatory Planning
- Tutorial: Engagement Visualisation: Desktop Visualisation or Game Engine?

Companion Website

Geodesign, Urban Digital Twins, and Futures companion site   www.Geodesigndigitaltwins.com

This introduction will cover several contemporary development areas, including notions of futuring, working definitions of geodesign, data sources, UDTs, and gamification. Each chapter-end section of this book will feature a case study for practical applications. Chapter 1 will scan climate challenges, notions of UDTs, and citizen science, particularly volunteered geographic information (VGI). Chapter 2 will critically examine concepts of UDTs and chart City Information Models (CIMs) of urban environments and their benefits (Souza & Bueno, 2022; Kitchin et al., 2021). Chapter 3 features a discussion of the geodesign framework for engagement and participation and its relevance to UDTs (Paradis et al., 2013). This chapter examines the process of geodesign and a number of case studies. Chapter 4 provides 11 geodesign methods for UDTs applicable to readers' research questions and evaluates various cases and adopted approaches. This chapter will discuss scenario generation, backcasting, and co-created futuring methods for choices for people, models, and platforms. Chapter 5 critically accounts for the increasing 'Gamification,' in GIS and Urban Planning, which is the application of gaming techniques in real-world settings. This chapter will assess the use of extended digital games for urban scenarios and communities and evaluate these tools for their engagement potential. Finally, in the concluding phase, we will scrutinise socio-technical challenges in implementing geodesign and UDTs for future environments.

FUTURING

Often, we can discuss the future, predict, and plan. Such predictions and plans result from a process of 'futuring.' Futuring is a process that is often multidisciplinary, co-created, informal, or formal. Futuring explores things that are yet to come from social aspects, health, physical development, and our built and natural environment. Moreover, 'futuring' has been a long concern and is particularly important in a world of finite resources, expanding and ageing populations, health, inequality, and depleting ecosystems in that many organisations and individuals are planning where we are heading and examining solutions and mitigations to these grand challenges. Futuring is not singular, often, the process generates heterogeneous outcomes, and this book seeks to embed GIS within these scenarios for resilience and robustness. Definitions of urban futures are highly plural, but much work is directed towards projecting, modelling, and speculating possible outcomes across various timespans, scales, and geographic areas (Pettit et al., 2019). Urban futures incorporate different vocabularies and territories, including economies, governance, habitat, and politics (Barua & Jellis, 2018). Urban futures could consist of foresight studies or horizon scans, systems modelling, and anticipation studies for rigorous forecasting, pursuing desired futures, representational practices, or epistemological critiques as part of methodologies (Minkkinen, 2020). As The UN report *Planning for Sustainable Cities* states,

Future urban planning needs to take place within an understanding of the factors that are shaping the socio-spatial aspects of cities and the institutional structures which attempt to manage them. It also needs to recognize the significant demographic and environmental challenges that lie ahead and for which systems of urban management will have to plan.
(2009, p. 36)

Standard methods of futuring that attempt to create this inclusiveness include backcasting, scenarios, gaming, and world-building amongst other modes. For example, Buckminster-Fuller's World Game in 1964 was developed as an approach to governance and social issues, planetary citizenship, and the development of 'Whole Earth' perspectives from participants (Figure 0.4). The game utilised

Figure 0.4
Buckminster-Fuller, World Game workshop, New York, 1969. The World Game was Fuller's approach to changing participants' perceptions of 'whole' systems approaches and regeneration of the Earth's ecosystem. Image courtesy of the Buckminster-Fuller Institute.

data analysis, systems modelling, scenario building, computers, and information design, and it was first played in North America. Notably, the World Game sometimes involved using film and media to create a more immersive experience for players to foster collaboration (Stott, 2022). If the World Game were rereleased, it would undoubtedly utilise the latest multimedia immersive techniques such as visualisation domes or virtual labs. Furthermore, futuring is often embedded within urban planning 'visions' at the government and local levels. Therefore, examining visions of future urban settings can also yield important information regarding processes, spatial reasoning, and the socio-technical perspectives present in the 'vision' setting by stakeholders (Dunn & Cureton, 2020). However, such a technique often reveals unconscious bias, professional dogmas, and the sourcing of original ideas in re-appropriated media.

Often, future-generated media or future visions have an impact on contemporary society. Imaginative work gains traction and wide social acceptance. These future projections can be seen in speculation and dystopian receptions of the impact of smart cities, big data, and privacy concerns (Townsend, 2013). For example, Alphabet's Sidewalk Labs, Toronto, Ontario, and its vision for the quay raised ethical questions about using Internet of Things (IoT) devices for commercial data collection from inhabitants and visitors over community engagement and place-making. Sidewalk Toronto's development plans and vision generated large counter forces in local opposition to the scheme, even though Sidewalk delivered a large and diverse range of community consultation methods. Sidewalk Lab's development is one that is a wicked problem; "the definition of a wicked problem is the problem itself." Often, one wicked problem is a symptom of another, and they are unique (Rittel & Webber, 1973, p. 161). Wicked problems have developed in planning education. Moria Zellner and Scott Campbell argue that they can align with complex systems, a range of planning methodologies, and simulations to adapt to these qualities. However, complex systems are different from tangible GIS ones,

> **While complex systems thinking shows promise in tackling and taming challenging planning issues like climate change and flooding, this framework has not been actively integrated into mainstream planning education and practice.**
>
> (Zellner & Campbell, 2015, p. 459)

GIS science could arguably expand to include supportive methods, such as 'futuring methods,' which are non-standardised and dynamic. For example, a long-standing issue for smart cities could revolve around information and communication technology (ICT)-centric approaches versus social and technological adoption and behaviours revolving around the uptake of digital services. Such an example creates tensions between specialists and experts and public response to such systems and dynamics. As Johan Colding et al. state,

> **Designers of Smart Cities should strive to build redundancy into the options affecting people's daily affairs in the digital city and avoid situations where non-choice default technologies become too dominant.**
>
> (Colding et al., 2019, p. 517)

Futuring methods can offer a meaningful exchange to wicked problems in urban planning and provide foresight to likely impacts (Oomen et al., 2021, p. 2). Often, deploying methods of iteration and ideation can progress projects towards the desired basis, alongside 'storying' the urban context to build resilience as part of the dissemination. Iteration and storying can be applied in both digital and analogue and in open-ended ways. For example, ESRI ArcGIS StoryMaps demonstrates this method embedded within contemporary web-based GIS. The Institute of Public Administration,

University of Delaware, conducted a project exploring public participation in geographic information systems (PPGIS) (Figure 0.5). As the report states,

> *The potential use of GIS Story Maps to fulfil mandates for an open government and public involvement by incorporating the use of interactive digital engagement tools into participatory planning processes. Online, interactive techniques and mapping applications are ideal for fostering citizen engagement, providing meaningful context to complex topics and concepts, and empowering informed decision making.*
>
> (Pragg, 2015, p. 2)

This project is but one of many seeking to understand participatory and collaborative methods for deeper engagement in planning processes and futuring methods to explore the complexity of the built

Figure 0.5
University of Delaware Institute for Public Administration and Delaware Department of Transportation, Complete Communities Toolbox and GIS Story Maps, 2015. The planning toolbox aims to build local government capacity through community design tools, including infographics and GIS StoryMaps. https://www.completecommunitiesde.org/community-design-tools/designing/.

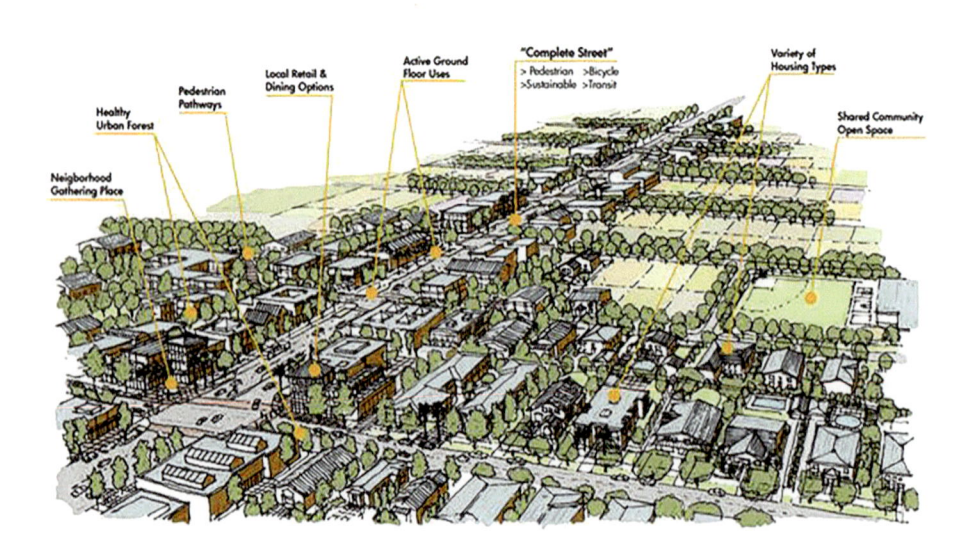

environment. As this book explores, these methods involve collaborative planning, not to deliver solutions but a range of tactics to respond to dynamic agents that cause problems through novel techniques such as gaming and others.

GEODESIGN

Motivations for this book revolve around charting the collaborative methods of working with GIS and their embedding in digital infrastructure. Central to futures is the field of geodesign, which is a framework for collaboration in imagining and understanding our future world. Carl Steinitz defined geodesign as "a set of questions and methods necessary to solve large, complicated, and significant design problems in the world…" (2012, p. 3) and is a process that merges geographical sciences and design professions in the built environment. Geodesign does not promote a purely digital dialogue nor seek to promote software products; instead, it seeks to develop consensus around planning issues through geographic analysis and suitability models for design and implementation. As Shannon McElvaney states, "Geodesign is an iterative design method that uses stakeholder input, geospatial modelling, impact simulations, and real-time feedback to facilitate holistic designs and smart decisions" (McElvaney & Walker, 2013). Many of these advancements above have sought to engage communities through digital transformation to shape future places. The collaborative framework for geodesign focuses on study areas and six modelling phases: representation, process, evaluation, change, impact, and decision. The fundamental ontology of geodesign is through the layering of physiographic and sociographic information building upon the work of Ian McHarg, particularly the summary of and collected studies found in **Design with Nature** (1969) (Figure 0.6) and the work of Patrick Geddes in developing regional analysis, which was inclusive of physical geography, economic activity, and anthropology, an interwoven pattern

GEODESIGN, URBAN DIGITAL TWINS AND FUTURES

Figure 0.6
Ian Mcharg. Delaware Basin, McHarg Papers, The Architectural Archives, University of Pennsylvania.

and relationship. These reframing approaches have been critical in establishing the role of GIS and its users. Indeed, as Michael Batty states,

> **The traditional model of design is quite static, but GIS is becoming much more a part of a global digital platform for any kind of spatial decision-making, and at any time and over any spatial scale, the activity of using GIS is highly dynamic. Conceptions of design have changed in a similar way and the new platforms that are emerging in the digital world are entirely consistent with the idea of geodesign.**
> (Batty, 2013, p. 1)

Geodesign seeks to improve spatial intelligence through opinion polling, often connecting with digital technology (Batty et al., 2024). As Kelleann Foster recognises, "Digital, spatial technology also plays an essential connector-role in fostering dialogue and understanding between the process participants: designers, scientists and the community" (Foster, 2016, p. 97). The connection ability also correlates with developments in smart cities, which were developed in 1994 with Amsterdam's 'digital city', later followed by Cisco Systems (2005) (Villa & Mitchell 2010) and IBM (2008) (Scuotto et al. 2016). ICT technologies, particularly GIS, have sought increased urban 'intelligence.' An example of geodesign methods applied to digital planning processes includes wind turbines in the Netherlands (Rafiee et al., 2018) and three scales of planning in Italy and France (Caglioni & Campagna, 2021; Campagna et al., 2019) (Figure 0.7). This book seeks to replicate geodesign collaborative methodologies, the power of GIS, and its contribution to urban intelligence, planning, and technology.

GEODESIGN, URBAN DIGITAL TWINS AND FUTURES

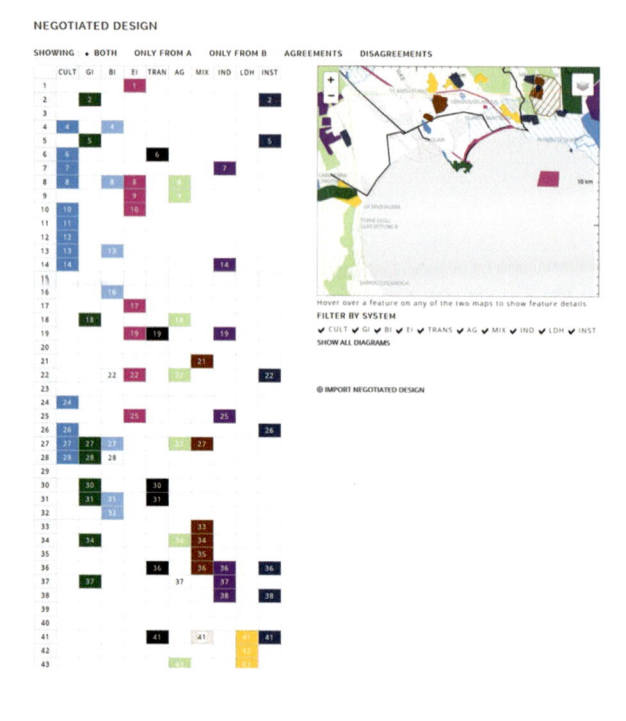

Figure 0.7
Michele Campagna, Metropolitan City of Cagliari (Italy) case study. Iteration for the International Geodesign Collaboration (IGC) project 2018. Images from the Geodesign Hub, a closed community platform for consensus planning designed by Hrishikesh Ballal. https://www.geodesignhub.com/.

DATA ACQUISITION, ACCESS TO DATA, AND OPEN DATA

There is abundant contemporary geographic data from government, non-governmental agencies, commercial companies, and VGI that highlight the pace of development and innovation from geographic data sources and aerial surveys (Figure 0.8). This data will have varying licence requirements, from restricted to commercial to open data. The frequency of this data, as well as its cost and updating, varies dramatically. Government mapping agencies are one of the first ports of call for developing projects, with sizeable satellite datasets, digital elevation models (DEM) or digital terrain models (DTM), and census information. These governmental datasets, coupled with third-party data, open data, or VGI such as LiDAR scans from iPhone or recreational drone flights can create complex data assemblages, making a selection and conceptual basis of the data framework often difficult for users when starting a project. An example of mixed data acquisition can be found in East Malaysia. The Sarawak Multimedia Authority created a governance information platform and surveyed multiple areas using ground-based vehicles, drones, and light

Figure 0.8
Aerial survey, Manhattan Island, New York City Cartographic. Maps, 1921. Lionel Pincus and Princess Firyal Map Division. Alamy Photo.

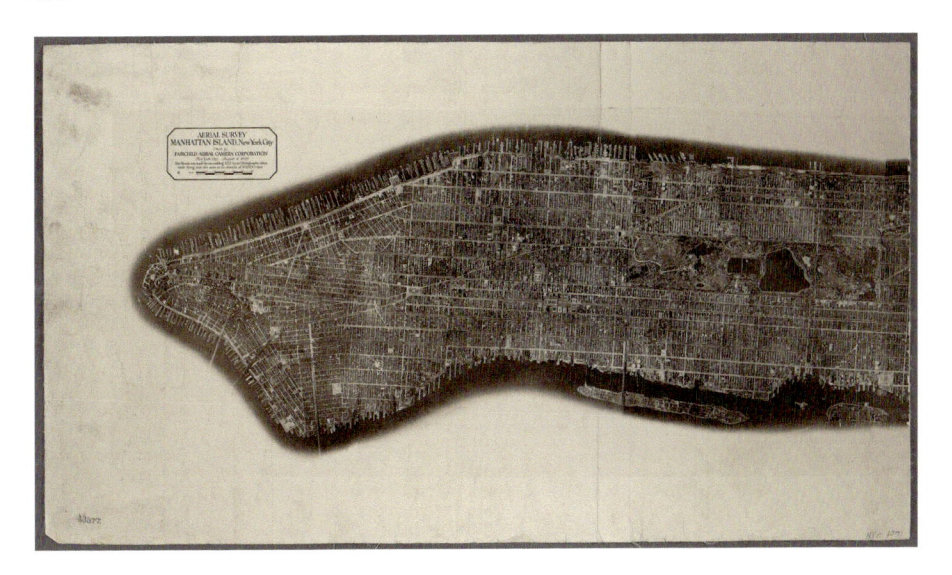

aircraft, followed by extensive modelling and extraction (Figure 0.9). Novel sensing may feature, for example, Bluesky International and the Leica City Mapper (Figure 0.10). What are the outputs? What resolution is required? Is there interoperability? What project outcomes are envisaged? How many iterations for a client? What analytical possibilities are possible? Can processes be automated? Can the data assist in timely project delivery? These are just a few basic questions that emerge when developing a brief. Rob Kitchin has argued that data is fundamental to knowledge production, yet little attention is paid to the ontological framing and assemblage surrounding its use (Kitchin, 2014, p. 184). Data ontologies are varied and have some specific discipline features. For example, 3D Virtual City Models use City GML, geographic markup language,

CityGML covers the geometrical, topological, and semantic aspects of 3D city models. The class taxonomy distinguishes between buildings and other man-made artifacts, vegetation objects, waterbodies, and transportation facilities like streets and railways. Spatial as well as semantic properties are structured in five consecutive levels of detail (LoD),

Figure 0.9
Sulaiman Bin Budin, Jabatan Tanah Dan Survey, 2019, Sarawak, Malaysia. Sarawak, a Malaysian state on Borneo, is being mapped and modelled into a comprehensive city information model using a variety of survey techniques and fused data, with reality and semantic models captured from UAV and mobile LiDAR mapping, oblique aerial imagery from a plane, open-source, and governmental data amongst many other fields. https://landsurvey.sarawak.gov.my/modules/web/pages.php?lang=bm&mod=webpage&sub=page&id=1621&menu_id=0&sub_id=326.

INTRODUCTION: GEODESIGN, URBAN DIGITAL TWINS, AND FUTURES

Figure 0.10
Bluesky International, 2023, Leica City Mapper, data product using hybrid oblique imaging and LiDAR solution. St Paul's Cathedral, London, UK.

where LOD0 defines a coarse regional model and the most detailed LOD4 comprises building interiors resp. indoor features.

(Kolbe et al., 2005, p. 884)

This data ontology has been designed for semantic interoperability to ISO standards and web usability. CityGML allows for rapid updating, classification, representation, and communication across multiple stakeholders (Biljecki et al., 2016). CityGML is one example of an ontological data model established through scholarship and long-term development by the open geospatial consortium or Building Smart International, among others. Other data model ontologies include GEOSci for geological features or LandInfra, a newer ontology for describing land and infrastructure features. Users of these products and other types should familiarise themselves with their inherent structures and standards (see table below). Data selection,

23

construction, framing, and assemblage are therefore essential in geodesign and urban future processes and are considered throughout each section and area of this book.

Example OGC Data Ontologies

LandInfra	https://www.ogc.org/standards/infragml
CityGML	https://www.ogc.org/standards/citygml
GEOSci	https://www.ogc.org/standards/geosciml

BuildingSMART International

Industry Foundation Classes (IFC)	https://technical.buildingsmart.org/standards/ifc/ifc-schema-specifications/
BIM (Building Information Modelling) Collaboration Format (BCF)	https://technical.buildingsmart.org/standards/bcf/

Example Standards

IGS Standards (GNSS Data)	https://www.igs.org/formats-and-standards/
ISO Unmanned Aircraft Systems	https://www.iso.org/committee/5336224.html

Open Data

Open Data Institute	https://theodi.org/

SMART CITIES AND DIGITAL TWINS

There have been many studies and interest in smart cities' urban analytics and informatics. Urban analytics uses "new forms of data in combination with computational approaches to gain insight into urban processes" (Singleton et al., 2018, p. 15). Urban informatics studies people utilising and applying data in the built environment. Constantine Kontokosta provides a working definition of urban informatics,

> *Urban informatics is the study of urban phenomena through a data science framework of urban sensing, data mining and integration, modelling and analysis, and visualization to generate new insights that*

simultaneously advance methods in computational science and address domain-specific urban challenges.
(Kontokosta, 2021, p. 383)

The focus on urban computing is part of paradigm changes to the major developed areas using big data, embedded sensors, and large ICT networks that are variegated and uneven, as evidenced by the Global Power City Index (GPCI), a mechanism evaluating major cities using six criteria of city function. The criteria are accessibility, economy, research and development, cultural interaction, liveability, and environment. This index, coupled and juxtaposed with UN studies of urbanisation, provides further evidence of the emergence of global 'super' cities. For supercities, technological development in city functionality has been defined through smart cities and IoT. According to Vito Albino, using a literature review, smart city definitions identified similar but multi-faceted values and divergent definitions, creating a 'fuzziness' and the inability for shared universality of the term (Albino et al., 2015). Indeed, the level of smartness beyond supercities is highly diverse. Technological 'smartness' and/or 'intelligence' are more likely to be studied at the regional and city levels due to the specificity of the components, interventions, level of development, implementation, and infrastructural capacity. Michael Batty (2018, p. 178) states that 'smartness' definitions have proved difficult because of the process aspect, and Nick Dunn and I (2019) have analysed the 'frictionless' vision aspect that many smart city projects present. Many of these wide-spanning issues are geographic ones or incorporate advanced GIS for urban management and delivery fused with new methods of designing.

We are at a stage of moving beyond static 2D representations of future forms but have data-informed models that are often time-based and generative. For example, Craig Taylor visualised new commuter flows using census data (Figure 0.11). The final rendered animation powerfully communicates the UK cities' public

Figure 0.11
Craig Taylor, Commuter Flowers, UK, ITO World, 2022. Utilising Houdini software, Taylor created commuter-like flowers from 2011 census data. By using origin and destination information, and routes via an open street map, Taylor created petal-like data visualisations of cities across the United Kingdom. Tutorials: https://mapzilla.co.uk/work/commuter-flowers.

transport networks. This project indicates many recent projects that use big urban data or real-time urban information to visualise urban conditions. GIS information has moved from static representations of vector and raster maps based on geodatabases held on local networks or single computer units to cloud-based repositories that can easily align other spatial datasets, allow dynamic updates, and increase access and dissemination of information on a multitude of platforms. Cartographic techniques and data visualisation are "the representation and presentation of data to facilitate understanding" (Kirk, 2020, p. 19). The increasing connectedness of 'smart cities' also means that GIS are increasingly important in urban governance. In particular, city 'dashboards' track performance. Changfeng Jing et al. suggest three forms of information design: operational, which describes measurements; analytical, which is diagnostic; and strategic dashboards, which are predictive. These information types have developed from early dashboard concepts in the 1990s (Jing et al., 2019, p. 2). Dashboards are one public-facing element of a complex smart city ecosystem that geospatial designers need to consider in development and construction, as seen in 51 World's model and dashboards for Beijing (Figure 0.12). One such growth area is the possibility of 'City Digital Twins,' 'Connected City Digital Twins,' or 'Smart City Digital Twins' as a development of smart cities which have captured the imagination in terms of urban futures but their relation to a specific place and reality. There are a variety of terms and definitions, but this book frames UDTs as an explicit working term defined in (Chapter 1).

Digital twins are intended as precise simulations of a physical asset, providing an "ability to simulate the behaviour of the system in digital form is a quantum leap in discovering and understanding emergent behaviour" (Grieves & Vickers, 2017, p. 90). A digital twin is thus a "mirror" that connects "real space, virtual space, the link for data flow from real space to virtual space, the link for information flow from virtual space to real space and virtual sub-spaces" (Grieves & Vickers, 2017, p. 85). This reflects the predictions of computer representations proposed by Gelernter (1993) in creating "mirror worlds," where detailed physical information becomes de-localised and data is held on a cloud server (Hudson-Smith, 2022). See, for example, the neural radiance field (NeRF), which creates a 3D representation based on 2D images, in this case, at the city scale (Xiangli et al., 2021; Song et al., 2023) (Figure 0.13). In another example, in the case of the City of Helsinki administration and its 3D open data model, which has been developing for decades, Zoan partnered with the city to make a virtual replica of the city for virtual reality immersive experiences in 2015 for community gathering, tourism, and concerts amongst many activities and is one of the first 'metaverses' or virtual world (KIRA–Digi, 2019) (Figure 0.14). Digital twins of urban and environmental settings have real developmental potential but vary drastically in their application and region.

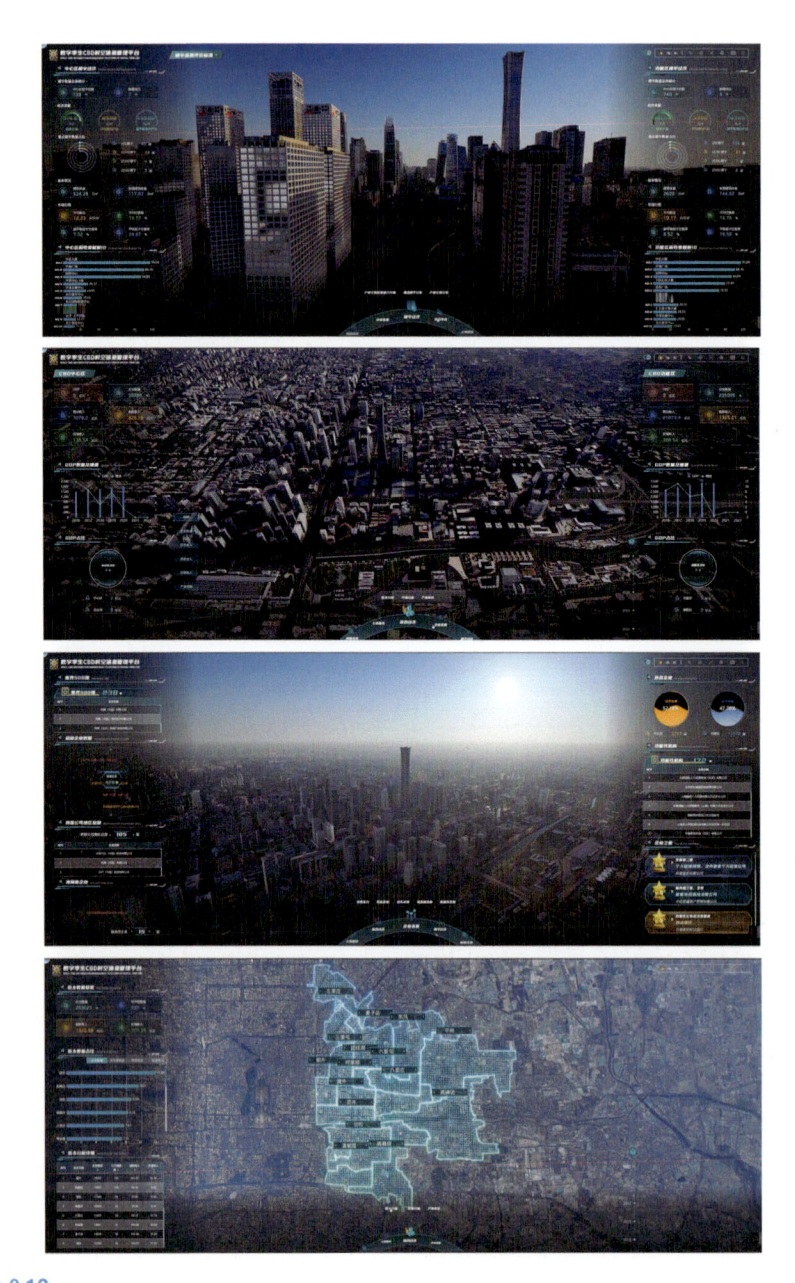

Figure 0.12
51 World, Digital Beijing, 2024. 51 World provides a platform as a service (PaS) for 3D interactions and city management using various sensing tools, including drones. 51 World has created over 100 cities in their workflow and provides full immersive experiences via Unreal Engine.

INTRODUCTION: GEODESIGN, URBAN DIGITAL TWINS, AND FUTURES

Figure 0.13
Song, K., Zeng, X., Ren, C., & Zhang, J. City on the Web, 2023. The existing neural radiance field (NeRF) renders small real-time scenes on a web platform. City-on-Web demonstrates a method for real-time rendering of large-scale scenes, though challenges remain with computational intensity and bandwidth.

Figure 0.14
Zoan, Virtual Helsinki, Finland, 2015. Zoan and the City of Helsinki created one of the first 'metaverses' and one of the biggest virtual reality concerts in the city, as well as utilising the model for virtual tourism.

29

GAMIFICATION

The range of visualisation tools, including GIS with urban planning, is part of a significant drive to support dialogue in planning processes via digital tools (Billger et al., 2017). Part of this contemporary dialogue involves using virtual environments, game design processes, and extended reality (XR devices) for immersion in these worlds or data fields. In essence, virtual interactions are anchored to reality. In an urban design and planning context, these mixed-reality systems are intended to increase public consultation, deliver efficiency, and improve decision-making processes (Sanchez-Sepulveda et al., 2019). These drivers have a much longer history in terms of the work of Cedric Price's Fun Palace, in which a flexible frame, therefore, can feature a multiplicity of actors and be programmed for many purposes; architecture is gamified for social utility (Figure 0.15). Price's view of technology for social purpose was also evident in collaboration with Gordon Pask for 'Techno Trees' for data information exchange in Kawasaki, Japan, in which domes are networked "for the revitalization

Figure 0.15
Cedric Price, model of Fun Palace project, 1961. RIBA Collection.

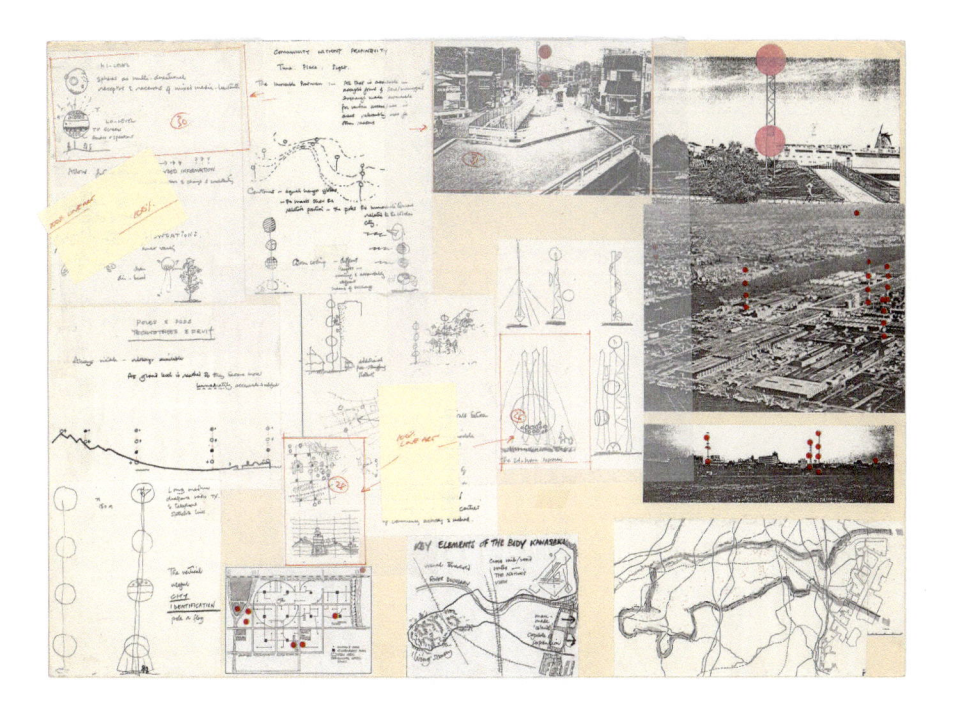

INTRODUCTION: GEODESIGN, URBAN DIGITAL TWINS, AND FUTURES

Figure 0.16
Cedric Price and Gordon Pask, presentation panels of the Kawasaki Project, 1986 montage 60 x 85 cm. Cedric Price Funds, Canadian Centre for Architecture © CCA.

of Kawasaki," to take it from a "long-established major industrial city" to an "information-intensive and humanistic city." Arguably Price and Pask's project is a prototype of UDTs (Palate et al., n.d.) (Figure 0.16).

ESRI's NFrames Photogrammetry software, which processes nadir and oblique imagery to create 3D meshes, means that simulative realities found in games are replicable to real-world environments. The Block-by-Block project began in 2012, a collaboration between Mojang, Microsoft, and UN-Habitat; it targeted non-traditional demographics who were engaged using a methodology which included the computer game Minecraft, in which users co-construct their world, design elements, and collaborate with professionals and communicating these decisions in mixed reality. Block-by-Block is an example of more meaningful community engagement and side-line issues of data accessibility and game architecture issues through an open platform and model creation from scratch (Chapter 4).

GIS and the fusion with BIM allow virtual reality applications of large-scale urban centres and regions (GeoBIM). The integration of geo-information is an essential supportive value structure for XR, and a large-scale study of European application for the EuroSDR Geo-BIM project by Noardo et al. (2020) has discussed the challenges in the regulatory environment, data conversion, and diversified user needs. This simulative possibility for multiscale planning information is more commonly extended to public consultations and communities through augmented reality (AR), as Devisch et al. (2016) show 'gamification' for urban planning. 'Gamification' is game design elements applied to non-game contexts (Deterding et al., 2011). However, urban planning has a much longer history of developing strategies and tactical (gamified) arrangements in analogue, and these aspects then became a rich trope in computer games history, as seen in Micropolis designed by Will Wright, the urban planning simulator, later becoming SimCity (Maxis, 1989; Devisch, 2008). City-building games such as SimCity are limited in their gamified contexts compared to real-world planning issues, but scenario-based games are valuable methods and tasks to undertake (Bereitschaft, 2016) (Chapter 5).

Many software development companies have sought to develop realism through game engines, creating 'real' or twinned environments that output to XR (Milgram et al., 1995). The communication of the built environment has been a pivotal force in developing virtual, mixed, and AR applications and experiential drivers to mitigate the complexity of the urban and construction processes. This book will consider gamified options, explore various software workflows, and consider audiences and engagement with XR to address one potential aspect of future UDTs.

SUMMARY

From futuring, geodesign, UDTs, and gamification, we can see examples of increasing applications of digital tools and the convergence of fields and disciplines. This introduction has covered some of the working definitions and fields of enquiry that the authors mark as convergence points for new urban technology and GIS paradigms. Readers seeking additional insight should follow the material cited in full for a more in-depth discussion. This book has also been informed by two key texts. Michael Batty's *The Computable City* (2024) and Li Wan, Timea Nochta, Junqing Tang, and Jennifer Schooling's book, *Digital Twins for Smart Cities: Conceptualisation, Challenges, and Practices* (2023).

Klaus Schwab forecasts and marks the Fourth Industrial Revolution as a system of transformation across countries and industries, with breakthroughs in "artificial intelligence, robotics, the Internet of Things, autonomous vehicles, 3-D printing, nanotechnology, biotechnology, materials science, energy storage, and quantum computing" (Schwab, 2016). The convergence and paradigm shift explored here is in our built environment and is evidenced and supported by various technological capabilities, cases, and projects we will explore. A diverse and developed framework and toolbox is presented in this book which offers new and or efficient methods of dialogue and engagement in the planning and design of places. Examples of important engagement activities appear in the work of 300,000 km/s showing night time visualisations for mapping Barcelona's characteristics and night activities (Figure 0.17). In another example, MVRDV's masterplan created an open tabula rasa for residents to shape spaces in by their own means (Figure 0.18). The areas of convergence charted by this book and the paradigm shifts occurring in the science of cities are the first publications of the author's knowledge that bring these critical developments together in one volume. In the silent science fiction dystopian film Metropolis, directed by Fritz Lang, an industrial-technological city is presented, which deals with society's mechanisation and class structures (Figure 0.19). Metropolis is a poignant vision and example of the future city, and the chapters presented here are mapped against the need to address complex future environments, build resilient futures, and reflect on the urban forms we want and those we do not.

GEODESIGN, URBAN DIGITAL TWINS AND FUTURES

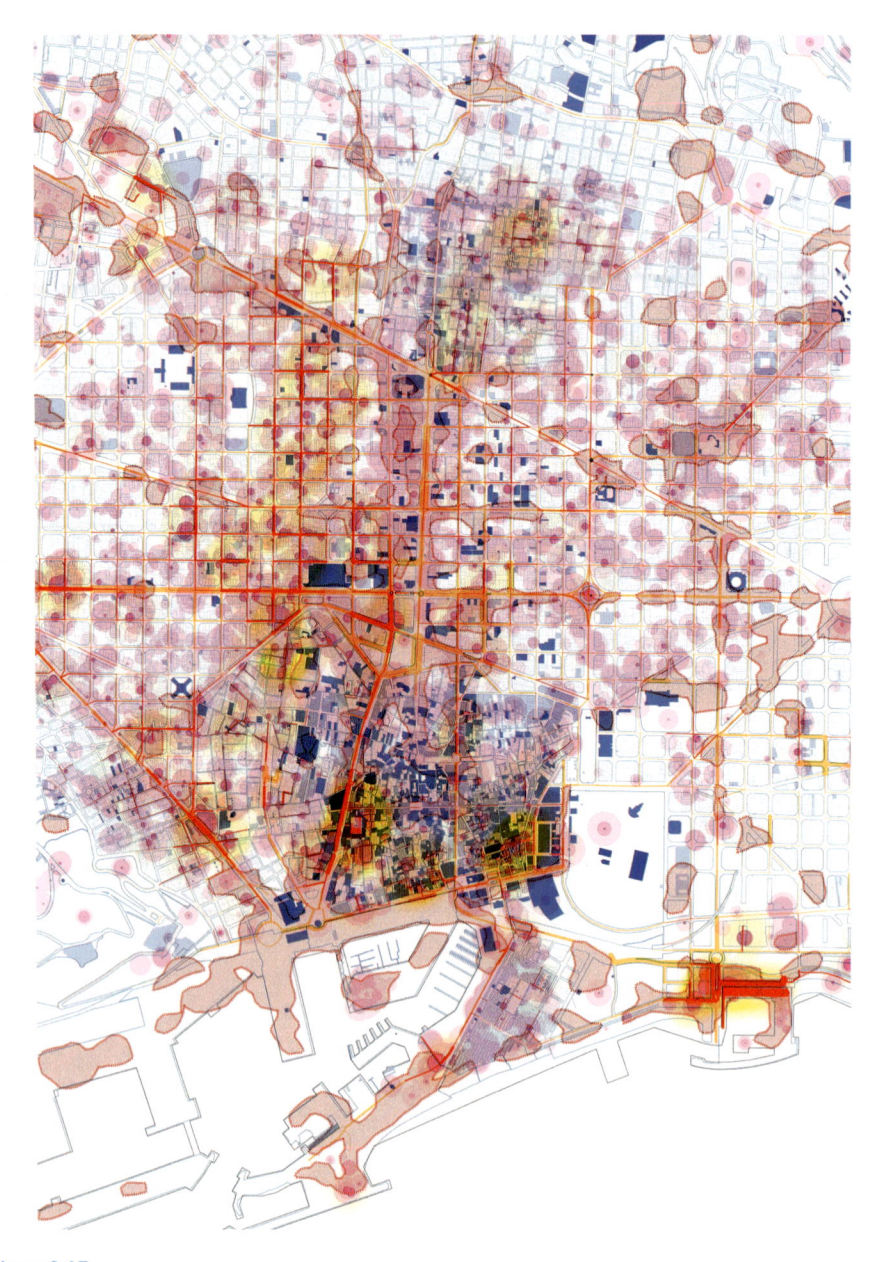

Figure 0.17
300,000 km/s, Barcelona Plaça Catalunya – Gran Via, 2018. Report on Barcelona's nightscape (p. 89 of the Atlas). https://300000kms.net/case_study/report-on-barcelonas-nightscape/. 300,000 km/s mapped the nighttime values and characteristics in order to map clusters of overlapping activity with daytime operations and areas for project development.

INTRODUCTION: GEODESIGN, URBAN DIGITAL TWINS, AND FUTURES

Figure 0.18
MVRDV, Almere Oosterwold, Netherlands, 2011.
'Do-it-yourself urbanism,' a free and adaptive scheme for residents to shape. Klaas Hofman, Chiara Quinzii, Mick van Gemert, Sara Bjelke, Jonathan Telkamp, Maarten Haspels, and Wing Yun. https://www.mvrdv.com/projects/32/almere-oosterwold.

Figure 0.19
A poster for Fritz Lang's 1927 film 'Metropolis' starring Brigitte Helm. (Photo by Movie Poster Image Art/Getty Images).

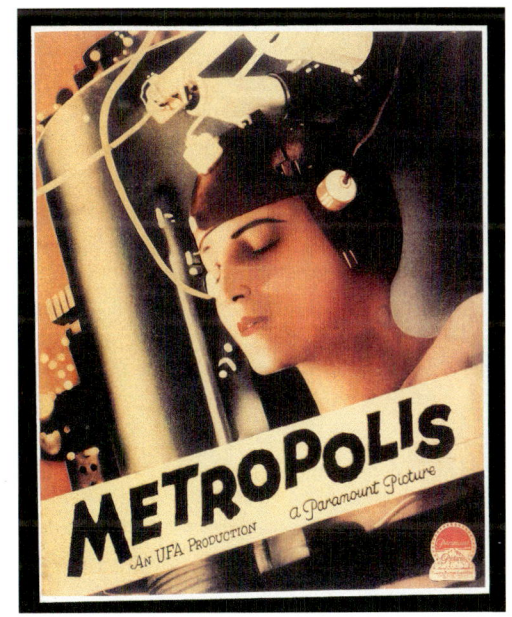

http://www.geodesigndigitaltwins.com/

35

REFERENCES

Albino, V., Berardi, U., Dangelico, R.M. (2015). Smart cities: Definitions, dimensions, performance, and initiatives. *Journal of Urban Technology*, 22(1), 3–21. https://doi.org/10.1080/10630732.2014.942092.

Barua, M., & Jellis, T. (2018). *Vocabularies for urban futures: Critical reflections*. British Academy. https://www.thebritishacademy.ac.uk/sites/default/files/urban-futures-vocabularies-urban-futures-critical-reflections.pdf.

Batty, M. (2013). Defining geodesign (=GIS + Design?). *Environment and Planning B: Planning and Design*, 40(1), 1–2. https://doi.org/10.1068/b4001ed.

Batty, M. (2018). *Inventing future cities*. MIT Press.

Batty, M. (2024). *The computable city*. MIT Press.

Batty, M., Shao, T., & Lopane, F. (2024). Participatory design based on opinion pooling. UCL CASA working paper 238. https://www.ucl.ac.uk/bartlett/casa/publications/2024/may/casa-working-paper-238.

Bereitschaft, B. (2016). Gods of the city? Reflecting on city building games as an early introduction to urban systems. *Journal of Geography*, 115(2), 51–60. https://doi.org/10.1080/00221341.2015.1070366.

Biljecki, F., Ledoux, H., & Stoter, J. (2016). Generation of multi-lod 3D city models in CityGML with the procedural modelling engine Random3Dcity. *ISPRS Annals of the Photogrammetry, Remote Sensing and Spatial Information Sciences*, IV-4/W1, 51–59. https://doi.org/10.5194/isprs-annals-IV-4-W1-51-2016.

Billger, M., Thuvander, L., & Wästberg, B.S. (2017). In search of visualization challenges: The development and implementation of visualization tools for supporting dialogue in urban planning processes. *Environment and Planning B: Urban Analytics and City Science*, 44(6), 1012–1035. https://doi.org/10.1177/0265813516657341.

Caglioni, M., & Campagna, M. (2021). Geodesign for collaborative spatial planning: Three case studies at different scales. In E. Garbolino, & C. Voiron-Canicio (Eds.), *Ecosystem and territorial resilience* (pp. 323–345). Elsevier. https://doi.org/10.1016/B978-0-12-818215-4.00012-2.

Campagna, M., Cocco, C., & di Cesare, E. (2019). New scenarios for the metropolitan city of Cagliari, Sardinia, Italy. In T. Fisher, B. Orland, & C. Steiitz (Eds.), *The international geodesign collaboration: Changing geography by design* (pp. 46–48). ESRI Press.

Chrisman, N. (2006). *Charting the unknown: How computer mapping at Harvard became GIS*. ESRI Press.

Colding, J., Barthel, S., & Sörqvist, P. (2019). Wicked problems of smart cities. *Smart Cities*, 2(4), 512–521. https://doi.org/10.3390/smartcities2040031.

Devisch, O. (2008). Should planners start playing computer games? Arguments from SimCity and second life. *Planning Theory & Practice*, 9(2), 209–226. https://doi.org/10.1080/14649350802042231.

Devisch, O., Poplin, A., & Sofronie, S. (2016). The gamification of civic participation: Two experiments in improving the skills of citizens to reflect collectively on spatial issues. *Journal of Urban Technology*, 23(2), 81–102. https://doi.org/10.1080/10630732.2015.1102419.

Dunn, N., & Cureton, P. (2019). Frictionless futures: The vision of smartness and the occlusion of alternatives. In S.M. Figueiredo, S. Krishnamurthy, & T. Schroeder (Eds.), *Architecture and the smart city* (pp. 17–28). (Critiques). Routledge.

Dunn, N., & Cureton, P. (2020). *Future cities: A visual guide*. Bloomsbury Publishing.

Foster, K. (2016). Geodesign parsed: Placing it within the rubric of recognized design theories. *Landscape and Urban Planning*, 156, 92–100. https://doi.org/10.1016/j.landurbplan.2016.06.017.

Gelernter, D. (1993). *Mirror worlds or the day software puts the universe in a Shoebox. How it will happen and what it will mean*. Oxford University Press.

Grieves, M., & Vickers, J. (2017). Digital twin: Mitigating unpredictable, undesirable emergent behaviour in complex systems. In F.-J. Kahlen, S. Flumerfelt, & A. Alves (Eds.), *Transdisciplinary perspectives on complex systems: New findings and approaches* (pp. 85–113). Springer International Publishing. https://doi.org/10.1007/978-3-319-38756-7_4.

Hall, P., & Tewdwr-Jones, M. (2020). Urban and regional planning (6th ed). Routledge.

Harrison, J., Galland, D., & Tewdwr-Jones, M. (2020). Regional planning is dead: Long live planning regional futures. *Regional Studies*, 1–13. https://doi.org/10.1080/00343404.2020.1750580.

Hessler, J. (2009). How to map a sandwich: Surfaces, topological existence theorems and the changing nature of modern thematic cartography, 1966–1972. ALA Map and Geography Round Table. Available electronically from https://hdl.handle.net/1969.1/129188.

Hudson-Smith, A. (2022). Incoming metaverses: Digital mirrors for urban planning. *Urban Planning*, 7(2), 343–354, ISSN 2183-7635.

Jing, C., Du, M., Li, S., & Liu, S. (2019). Geospatial dashboards for monitoring smart city performance. *Sustainability*, 11(20), 5648. https://doi.org/10.3390/su11205648.

KIRA–Digi. (2019). *The Kalasatama digital twins project digital twins project*. Ministry of the Environment.

Kirk, A. (2020). *Visualizing data*. Sage Publications.

Kitchin, R. (2014). *The data revolution: Big data, open data, data infrastructures & their consequences*. SAGE Publications Ltd.

Kitchin, R., Young, G.W. & Dawkins, O. (2021). Planning and 3D spatial media: Progress, prospects, and the knowledge and experiences of local government planners in Ireland. *Planning Theory & Practice*, 22(3), 349–367. https://doi.org/10.1080/14649357.2021.1921832.

Kolbe, T.H., Gröger, G., & Plümer, L. (2005). CityGML: Interoperable access to 3D city models. In P. van Oosterom, S. Zlatanova, & E.M. Fendel (Eds.), *Geo-information for disaster management*. Springer. https://doi.org./10.1007/3-540-27468-5_63.

Kontokosta, C.E. (2021). Urban informatics in the science and practice of planning. *Journal of Planning Education and Research*, 41(4), 382–395. https://doi.org/10.1177/0739456X18793716.

Maxis. (1989). SimCity [Digital game]. Maxis.

McElvaney, S., & Walker, D. (2013). Geodesign – Strategies for urban planning. In *American Planning Association national planning conference*, Chicago, IL, April 14.

McHarg, I.L. (1969). *Design with nature*. John Wiley.

Milgram, P., Takemura, H., Utsumi, A., & Kishino, F. (1995). Augmented reality: A class of displays on the reality-virtuality continuum. In *Proceedings of SPIE 2351, telemanipulator and telepresence technologies* (21st December). https://doi.org/10.1117/12.197321.

Minkkinen, M. (2020). Theories in futures studies: Examining the theory base of the futures field in light of survey results. *World Futures Review*, 12(1), 12–25. https://doi.org/10.1177/1946756719887717.

Noardo, F., Behan, A., Bucher, B., Devys, E., Ellul, C., Harrie, L., & Stoter, J. (2020, July 16). *EuroSDR GeoBIM project - Integrated workflow using GeoBIM information for building permit process (version 1)*. Zenodo. https://doi.org/10.5281/zenodo.3948493.

Oomen, J., Hoffman, J., & Hajer, M.A. (2021). Techniques of futuring: On how imagined futures become socially performative. *European Journal of Social Theory*. https://doi.org/10.1177/1368431020988826.

Page, J., Mörtberg, U., Destouni, G., Ferreira, C., Näsström, H., & Kalantari, Z. (2020). Open-source planning support system for sustainable regional planning: A case study of Stockholm County, Sweden. *Environment and Planning B: Urban Analytics and City Science*, 47(8), 1508–1523. https://doi.org/10.1177/2399808320919769.

Palate, S., Gerlinde, V., & Temizel, E. (n.d.). In the scale of... a reference number, a cabinet, a postbox. (Accessed 02/02/2024) https://www.cca.qc.ca/

en/articles/85197/in-the-scale-of-a-reference-number-a-cabinet-a-postbox.

Paradis, T., Treml, M., & Manone, M. (2013). Geodesign meets curriculum design: Integrating geodesign approaches into undergraduate programs. *Journal of Urbanism*, 6(3), 274–301. https://doi.org/10.1080/17549175.2013.788054.

Pettit, C.J., Hawken, S., Ticzon, C., Leao, S.Z., Afrooz, A.E., Lieske, S.N., Canfield, T., Ballal, H., & Steinitz, C. (2019). Breaking down the silos through geodesign – Envisioning Sydney's urban future. *Environment and Planning B: Urban Analytics and City Science*, 46(8), 1387–1404. https://doi.org/10.1177/2399808318812887.

Pragg, S. (2015). *Planning for complete communities in delaware strategic promotion and dissemination of the online delaware complete communities planning toolbox.* Institute for Public Administration School of Public Policy & Administration, College of Arts & Sciences, University of Delaware. https://www.complete-communitiesde.org/.

Rafiee, A., Van der Male, P., Eduardo Dias, E., & Scholten, H. (2018). Interactive 3D geodesign tool for multidisciplinary wind turbine planning. *Journal of Environmental Management*, 205, 107–124, ISSN 0301-4797. https://doi.org/10.1016/j.jenvman.2017.09.042.

Rittel, H.W.J., & Webber, M.M. (1973). Dilemmas in a general theory of planning. *Policy Sciences*, 4, 155–169.

Sanchez-Sepulveda, M., Fonseca, D., Franquesa, J., Redondo, E. (2019). Virtual interactive innovations applied for digital urban transformations. *Mixed approach, Future Generation Computer Systems*, 91, 371–381. ISSN 0167-739X. https://doi.org/10.1016/j.future.2018.08.016.

Schwab, k. (2016). The fourth industrial revolution: What it means, how to respond. https://www.weforum.org/agenda/2016/01/the-fourth-industrial-revolution-what-it-means-and-how-to-respond/.

Scuotto, V., Ferraris, A. & Bresciani, S. (2016), "Internet of Things: Applications and challenges in smart cities: a case study of IBM smart city projects", *Business Process Management Journal*, 22 No. 2, 357–367. https://doi.org/10.1108/BPMJ-05-2015-0074

Sebastian Deterding, Dan Dixon, Rilla Khaled, and Lennart Nacke. 2011. From game design elements to gamefulness: defining "gamification". In Proceedings of the 15th International Academic MindTrek Conference: Envisioning Future Media Environments (MindTrek '11). Association for Computing Machinery, New York, NY, USA, 9–15. https://doi.org/10.1145/2181037.2181040

Singleton, A., Spielman, S., & Folch, D. (2018). *Urban analytics.* Sage.

Song, K., Zeng, X., Ren, C., & Zhang, J. (2023). City-on-web: Real-time neural rendering of large-scale scenes on the web. ArXiv:2312.16457.

Souza, L., & Bueno, C. (2022). City Information Modelling as a support decision tool for planning and management of cities: A systematic literature review and bibliometric analysis. *Building and Environment*, 207, Part A, 108403, ISSN 0360-1323. https://doi.org/10.1016/j.buildenv.2021.108403.

Stott, T. (2022). *Buckminster Fuller's world game and its legacy*. Routledge.

Steinitz, C. (2012). *A framework for geodesign: Changing geography by design presents the key concepts, history, and methodology of geodesign*. ESRI Press.

Townsend, A.M. (2013). Smart cities: Big data, civic hackers, and the quest for a New Utopia (1st ed.). W.W. Norton & Company.

UN-Habitat. (2009). Planning sustainable cities: Global report on human settlements. https://unhabitat.org/planning-sustainable-cities-global-report-on-human-settlements-2009.

U.N. Habitat. (2022). *World cities report 2022*. https://unhabitat.org/wcr/.

Villa, N. & Mitchell, S. (2010) *Connecting Cities Achieving Sustainability Through Innovation*, White Paper (Accessed 10/05/2023) https://www.cisco.com/c/dam/en_us/about/ac79/docs/innov/Connecting_Cities_Sustainability_Through_Innovation_IBSG_1021FINAL.pdf

Wan, L., Nochta, T., Tang, J., & Schooling, J. (2023). *Digital twins for smart cities: Conceptualisation, challenges and practices*. ICE Publishing.

Zellner, M., & Campbell, S.D. (2015). Planning for deep-rooted problems: What can we learn from aligning complex systems and wicked problems? *Planning Theory & Practice*, 16(4), 457–478. https://doi.org/10.1080/14649357.2015.1084360.

Xiangli, Y., Xu, L., Pan, X., Zhao, N., Rao, A., Theobalt, C., Dai, B., & Lin, D. (2021). BungeeNeRF: Progressive neural radiance field for extreme multi-scale scene rendering. ArXiv:2112.05504.

CHAPTER 1

DEFINING THE LIVING LAB

INTRODUCTION

Of the United Nations (UN) Sustainable Development Goals (SDGs), SDG 11 – 'Making cities and human settlements inclusive, safe, resilient and sustainable' is significant in relation to geodesign and urban digital twins (UDTs) (Figure 1.1). Living Labs are an approach to establishing ecosystems for innovation and experimentation. In defining the parameters of a living lab for geodesign and UDTs, SDGs are critical for this framing, and this chapter sets out the various research challenges, data, and environments. This chapter discusses the policy environment and climatic challenges relevant to establishing UDTs. It will then discuss the UDT ecosystems by establishing a technological readiness level (TRL) for benchmarking, discussing modelling requirements, and discussing intra-disciplinary teams necessary for UDT implementation. Following this, bottom-up approaches for UDTs are offered. Design fiction is discussed as a design research method to explore Technological Readiness Levels 2–4 through a prototype for citizen-sensed data at scale. Citizen-sensed data, in particular, volunteered geographic information (VGI), is critically important to addressing global urbanisation goals and strategies and has a high potential for image generation and feedback to national and governmental UDT initiatives. VGI is discussed as a validating bottom-up approach for centralised governmental datasets and providing a near-real-time feedback loop to large-scale UDTs.

With over half of the global population residing in urban areas (55%), this is expected to rise to 70% by 2050. A raft of urban issues is present, including access to public transportation, urban sprawl, air pollution, and limited open public spaces. In particular, the SDG 1.3.2 Indicator seeks to deliver a "proportion of cities with a direct participation structure of civil society in urban planning and management that operates regularly and democratically" (U.N., 2022.). This global policy drive is supported by various mechanisms, including a national sample of cities (NSC) comprising 200 cities, which

Adverse impacts from human-caused climate change will continue to intensify

a) Observed widespread and substantial impacts and related losses and damages attributed to climate change

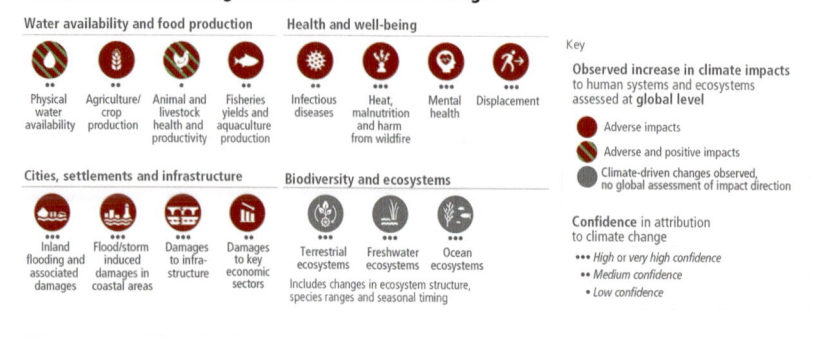

Water availability and food production
- Physical water availability
- Agriculture/crop production
- Animal and livestock health and productivity
- Fisheries yields and aquaculture production

Health and well-being
- Infectious diseases
- Heat, malnutrition and harm from wildfire
- Mental health
- Displacement

Cities, settlements and infrastructure
- Inland flooding and associated damages
- Flood/storm induced damages in coastal areas
- Damages to infrastructure
- Damages to key economic sectors

Biodiversity and ecosystems
- Terrestrial ecosystems
- Freshwater ecosystems
- Ocean ecosystems

Includes changes in ecosystem structure, species ranges and seasonal timing

Key
Observed increase in climate impacts to human systems and ecosystems assessed at **global level**
- Adverse impacts
- Adverse and positive impacts
- Climate-driven changes observed, no global assessment of impact direction

Confidence in attribution to climate change
- ••• High or very high confidence
- •• Medium confidence
- • Low confidence

b) Impacts are driven by changes in multiple physical climate conditions, which are increasingly attributed to human influence

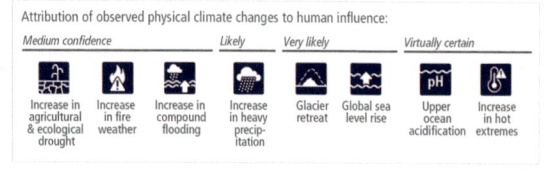

Attribution of observed physical climate changes to human influence:

Medium confidence	Likely	Very likely	Virtually certain
Increase in agricultural & ecological drought	Increase in fire weather	Increase in compound flooding	Increase in heavy precipitation / Glacier retreat / Global sea level rise	Upper ocean acidification / Increase in hot extremes

c) The extent to which current and future generations will experience a hotter and different world depends on choices now and in the near-term

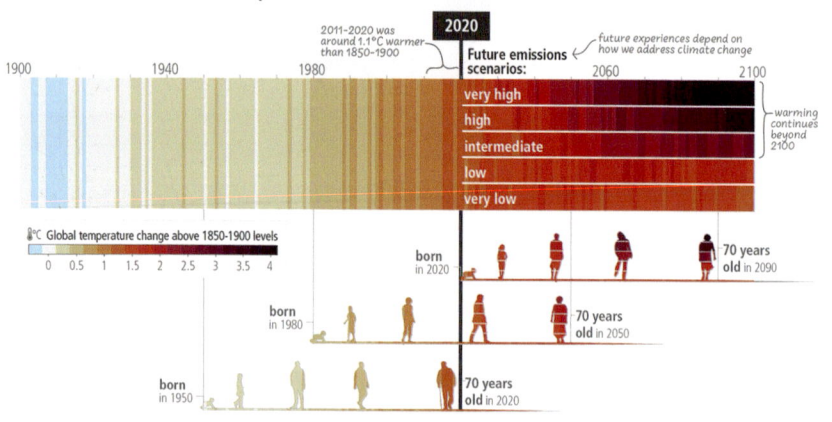

2011-2020 was around 1.1°C warmer than 1850-1900

2020

Future emissions scenarios:
- very high
- high
- intermediate
- low
- very low

future experiences depend on how we address climate change

warming continues beyond 2100

°C Global temperature change above 1850-1900 levels
0   0.5   1   1.5   2   2.5   3   3.5   4

born in 2020 — 70 years old in 2090
born in 1980 — 70 years old in 2050
born in 1950 — 70 years old in 2020

Figure 1.1

SPM.1: "(a) Climate change has caused widespread impacts, losses, and damages on human systems and altered terrestrial, freshwater, and ocean ecosystems worldwide. Mental health and displacement assessments have been conducted in some different regions. Confidence levels reflect the assessment of attribution of the observed impact of climate change. 'Changes in annual global surface temperatures are presented as "climate stripes," with future projections showing the human-caused long-term trends and continuing modulation by natural variability (represented here using observed levels of past natural variability). Colours on the generational icons correspond to the global surface temperature stripes for each year, with segments on future icons differentiating possible future experiences'" (IPCC, 2023).

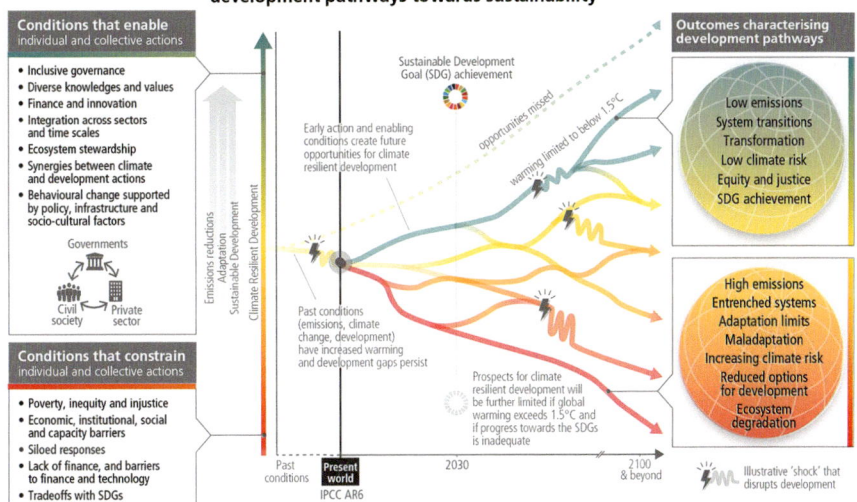

Figure 1.1 (Continued)
SPM.6: The illustrative development pathways (red to green) and associated outcomes (right panel) show that there is a rapidly narrowing window of opportunity to secure a liveable and sustainable future for all. Climate resilient development is the process of implementing greenhouse gas mitigation and adaptation measures to support sustainable development. Diverging pathways illustrate that interacting choices and actions made by diverse government, private sector and civil society actors can advance climate-resilient development, shift pathways towards sustainability, and enable lower emissions and adaptation. Diverse knowledge and values include cultural values, Indigenous Knowledge, local knowledge, and scientific knowledge. Climatic and non-climatic events, such as droughts, floods or pandemics, pose more severe shocks to pathways with lower climate-resilient development (red to yellow) than to pathways with higher climate-resilient development (green). (IPCC, 2023)

was achieved through the analysis of satellite imagery (Figure 1.2). Worldwide, spatial data will be collected in various ways to inform policy. In the EU case, the INSPIRE Directive addresses 34 themes for interoperable environmental applications. At the same time, the Intergovernmental Panel on Climate Change (IPCC, 2023), in the AR6 Synthesis report, states several factors with high confidence that are creating significant climate change (Figure 1.3). An extract of the factors is the following (Figure 1.4),

GEODESIGN, URBAN DIGITAL TWINS AND FUTURES

Figure 1.2
UN SDG 11. The UN defines goal Sustainable Development Goal 11 as making cities and human settlements inclusive, safe, resilient, and sustainable. Currently, 1.1 billion people live in slum-like conditions in cities. Projections state that 70% of the global population will live in cities by 2050.

Figure 1.3
The SDG 11 Explorer is a tool for easy exploration and analysis of land consumption. Users can access, explore, and compare land consumption trends. VISIT is a settlement footprint used to analyse urbanisation. The current collection provides access to WSF 2019, WSF 2015, and GUF 2011.

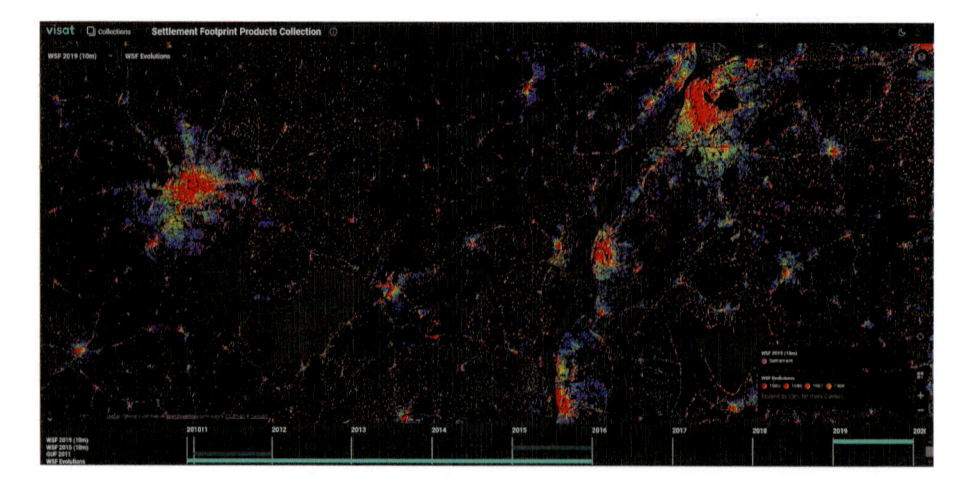

CHAPTER 1
DEFINING THE LIVING LAB

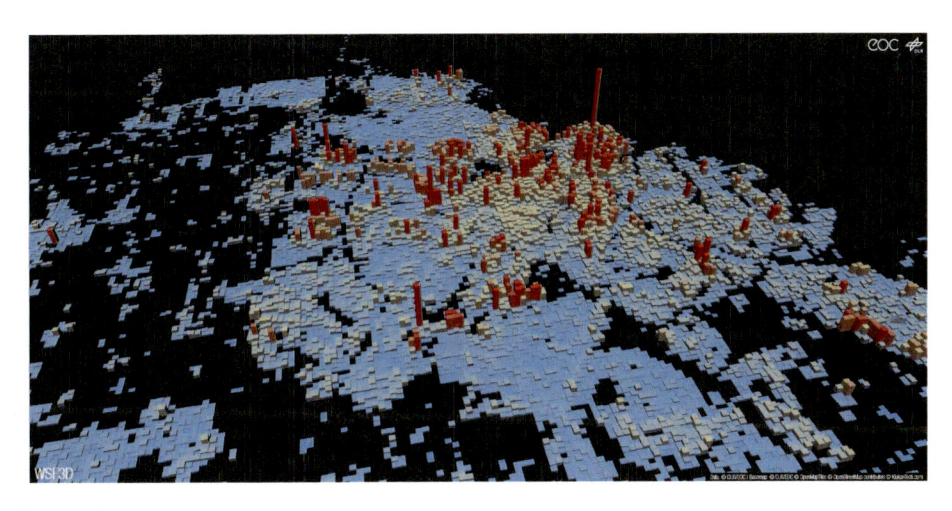

Figure 1.4
The World Settlement Footprint (WSF) 3D datasets provide detailed quantification of the average height, total volume, total area, and the fraction of buildings at 90m resolution at a global scale. It is generated using a modified version of the World Settlement Footprint human settlements mask derived from Sentinel-1 and Sentinel-2 satellite imagery in combination with digital elevation data and radar imagery collected by the TanDEM-X mission. The WSF shows a view of Malmo, Sweden.

A.2.6 Climate change has caused widespread adverse impacts and related losses and damages to nature and people that are unequally distributed across systems, regions and sectors. Economic damages from climate change have been detected in climate-exposed sectors, such as agriculture, forestry, fishery, energy, and tourism. Individual livelihoods have been affected through, for example, destruction of homes and infrastructure, and loss of property and income, human health and food security, with adverse effects on gender and social equity. (high confidence).

A.2.7 In urban areas, observed climate change has caused adverse impacts on human health, livelihoods and key infrastructure. Hot extremes have intensified in cities. Urban infrastructure, including transportation, water, sanitation and energy systems have been compromised by extreme and slow-onset events, with resulting economic losses, disruptions of services and negative impacts to well-being. Observed adverse impacts are concentrated amongst economically and socially marginalised urban residents. (high confidence).

(IPCC, 2023, p. 6)

There is a significant need for near-term change in leadership, decision-making, and implementation for climate-resilient development. Fundamentally, the motivation of the book is the underlying philosophy that the combination of collaborative frameworks for decision-making combined with technological apparatus through federated UDTs may address a number of these urgent challenges and mitigations. However, several research challenges exist regarding geographies, financing, data, and governance. Geodesign is a collaborative framework but has yet to be applied at scale to UDTs. UDTs are also an immature systems approach for the built environment. The potential for UDTs to address SDGs rests on the ability of the system to create future scenarios in which models and parameters are focused on SDG policy goals. As Asaf Tzachor et al. states,

> **The accuracy and precision of the representation that a digital twin provides for an object, which forms the basis of the decision-making support utility, depends on the available, accessible and compatible data.**
> (Tzachor et al., 2022, p. 824)

There is some convergence and demonstration of interoperable data with clear policy indicators. These relationships have strong potential; for example, SDGs have been embedded in the city of Melbourne, Australia, Digital Twin (Barzegar et al., 2021) (Figure 1.5). A framework for Melbourne's implementation in relation to SDGs is already in place for planning (Allan et al., 2024). The range of digital twin research projects by the Centre for Spatial Data Infrastructures and Land Administration, University of Melbourne, include aggregated data, a coastal flood platform, a real-time decision support tool, a pedestrian simulator, and planning development control, amongst other applied cases (Irajifar, 2022; Langenheim et al., 2022). Melbourne may be an anomaly in addressing top-down SDG goals and there is further work in this area (Bloch & Sacks, 2018). As Cecilia Wong states, "top-down effort alone is inadequate to serve the complex urban agenda […] critical local issues identified

CHAPTER 1
DEFINING THE LIVING LAB

Figure 1.5
Centre for Spatial Data Infrastructures and Land Administration, Fishermans Bend Digital Twin, University of Melbourne, 2022. Fisherman's Bend precinct covers 480 ha in the heart of Melbourne to visualise historic and future scenarios.

from the bottom-up process can be set against the top-down strategic goal…" (Rae & Wong, 2021, pp. 33–34). It is precisely the approach described by Wong that this chapter seeks to address.

Developing a digital twin system to match global policy ambitions and test at scale is a significant research challenge. In the issue of accessible and compatible data, in the spatial planning profession, mechanisms such as land use are defined as the use and functions of a city, authority, region, or via national coverage. This could be residential, commercial, industrial, transport network, vegetation, etc… The land use database will describe various terms of land and its socio-economic and ecological function through nomenclature. Land Cover is the surface description of the bio-physical characteristics of the spatial area under coverage through classification, see for example (Figure 1.6) of the Lake District, United Kingdom, by the UK Centre for Ecology & Hydrology (UKCEH). For example, CORINE is a harmonised dataset of land cover and land use for Europe of 44 classes, and ALUM is a land use and management classification for Australia. However, no unified land use classification across academia or other professions exists for decisions. Problems such as data sharing, different nomenclature, resolutions, scales, granular details, and infrastructure are often reported with attempts to establish a unified system. Satellite imagery has remained one of the mainstays of remote sensing,

49

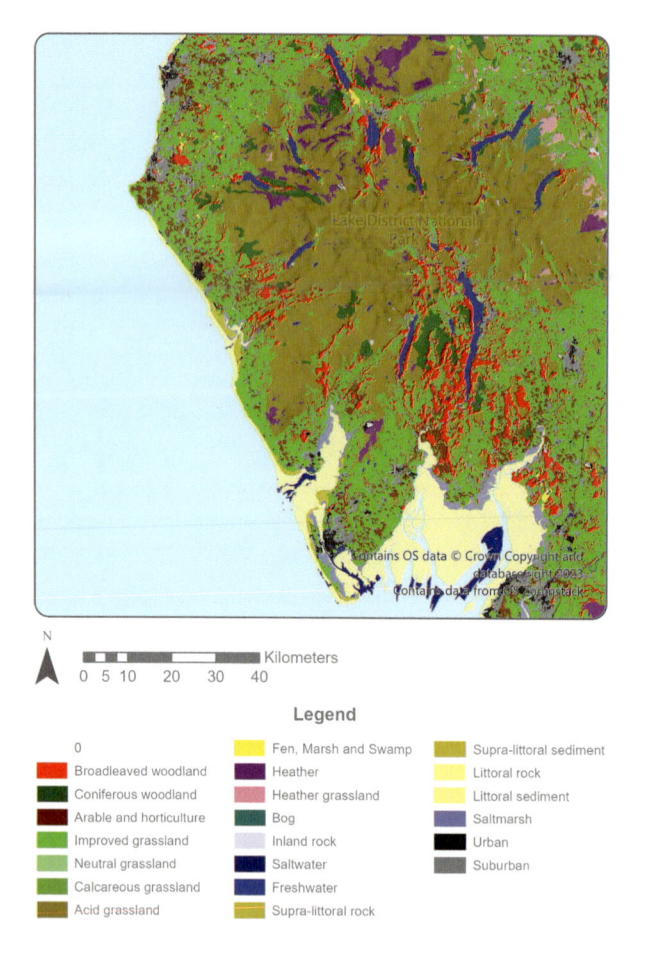

Figure 1.6
UKCEH, Integrated Hydrological DTM, Land Cover Maps, Lake District, United Kingdom, 2021 Land Cover Map. © Ordnance Survey Open Data.

from which land use classifications derive. However, other image sets derived from other sources are often used alongside various sensors, including mobile signalling and metrological and environmental devices. Thus, for UDTs taking the example of competing systems of land use, there are questions about the viability of decision-making systems, given the existing challenges and competing systems.

How do we approach the planning of cities, urban and rural areas, and regions for climate resilience? What role do various technological tools, systems, statutory planning instruments, and mechanisms play in shaping our ideas, relationships, and interactions with these places?

The study of futures for urban science is often incorporated and highly interdisciplinary, encompassing health and well-being, urbanisation, planning and growth, governance, and economic geography. Weiqi Zhou et al. synthesised five frameworks that call for a global framework of urban science to advance sustainability. The goals consist of:

1. **the joint social, ecological, and technological nature of urban systems**
2. **the role of disruptive actions in urban transformation**
3. **the capacity of urban systems to adapt**
4. **the interactive and dynamic complexity of urban spatial form**
5. **transboundary flows at multiple scales (Zhou et al., 2021, p. 1).**

Such synthesis work is vitally important, as is further work in providing applied cases of UDTs to reduce the considerable ambiguity of the term itself. This ambiguity can be reduced through urban futures research, which employs various methods focusing on projecting, modelling, and speculating possible outcomes across different timespans, scales, and geographic areas. Some standard methods used in futuring include backcasting, scenario planning, gaming, world-building, and urban prototyping (Chapter 4). For example, participants could use data analysis, systems modelling, scenario building, information design, and satellite imaging in educational settings and collaborative workshops to explore global citizenships (Stott, 2022). Geodesign is a collaborative framework for participation across design and natural sciences, which has been present for over 30 years (Chapter 3), and UDTs are the systems approach and technological elements for urban management (Chapter 2) (van der Aalst et al., 2021). The combination and overlap of these areas are critical in addressing the SDGs, particularly 11. GIS Science is a recurring theme in both the frameworks for collaboration and enabling technology. Various land use modelling applications are then applied using these baseline databases and classifications. This could be agent-based models (Cellular Automata) for land use changes or

GEODESIGN, URBAN DIGITAL TWINS AND FUTURES

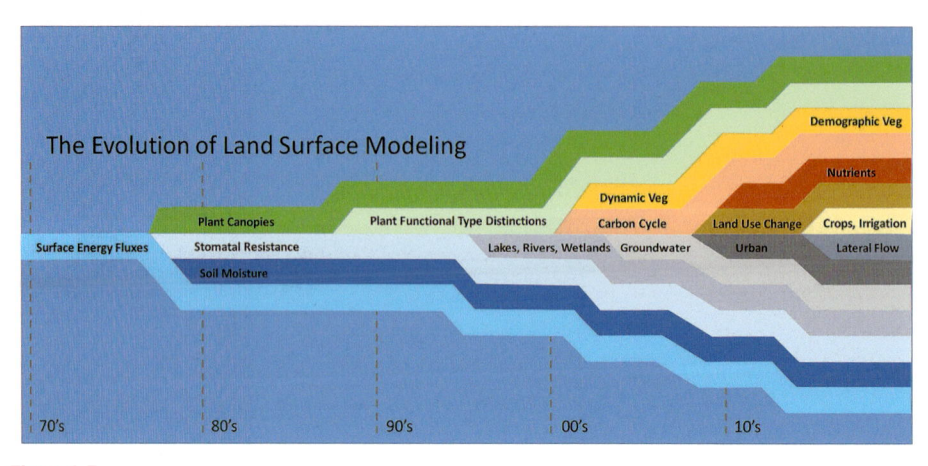

Figure 1.7
Rosie A. Fisher and Charles D. Koven, 2020. Land Surface Models. "A schematic depiction of the evolution of land surface model process representation through time, representing the emergence of different model components as commonly employed features of Earth system models" (Fisher & Koven, 2020, p. 2).

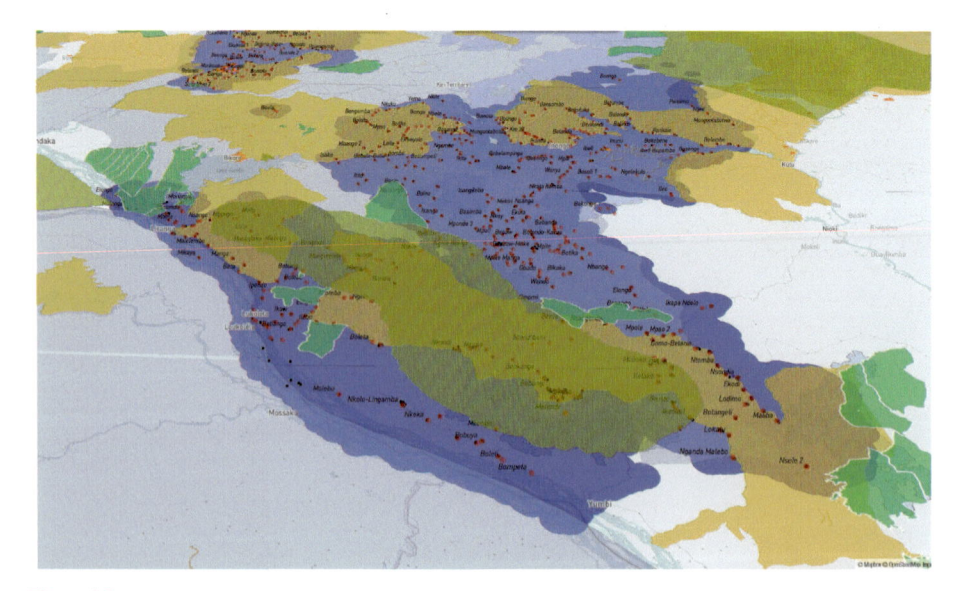

Figure 1.8
Mapping for Rights, RFUK, 2016. The atlas contains over 1,000 maps covering more than 9 million ha produced by rainforest communities through low-cost, easy-to-use mobile applications to understand the dynamics of forest use and harmful impacts and rights-based approaches to forestry management. https://www.mappingforrights.org/.

CHAPTER 1
DEFINING THE LIVING LAB

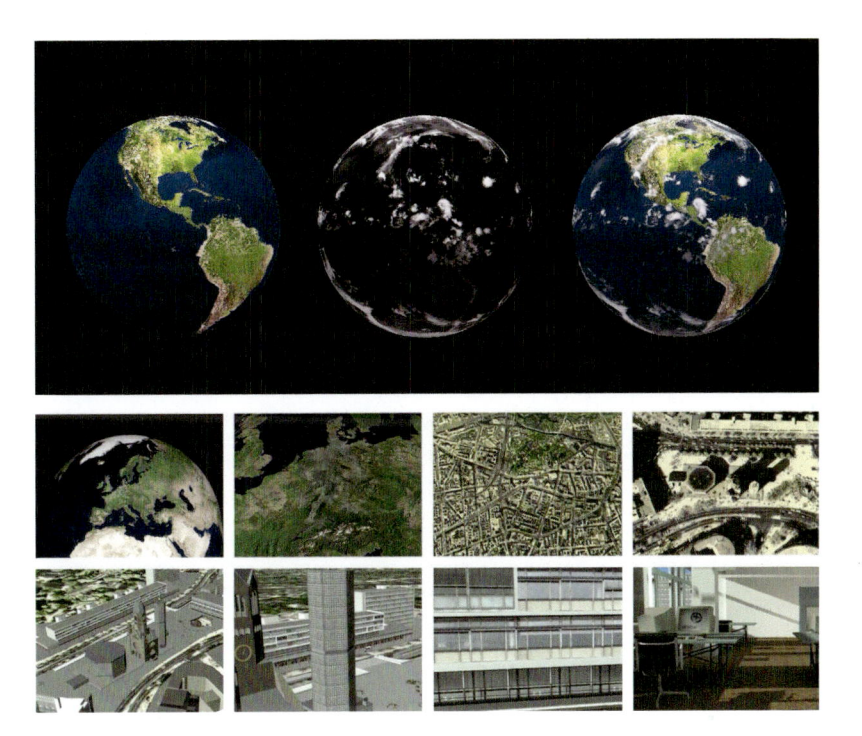

Figure 1.9
Art + Com, Terra Vision, 1994. Art + Com created a photorealistic virtual globe to which users could freely navigate at various scales. In order to navigate this data, Art + Com created a large sphere referencing the globe to pilot the planet, a 3D mouse to fly around, and a touch screen to interact with objects. Terravision was the first system to provide seamless navigation and visualisation in a massively large spatial data environment.

land surface models (LSMs) (Figure 1.7), which are common and highly varied and increasing in complexity. LSMs become critical devices, as in the case of Rainforest UK, through a strict methodological approach to defining land use and indigenous rights (Figure 1.8). Fundamentally, the range of geographic technologies, analytical abilities, and uses is fundamentally changing, and this change can be particularly highlighted in the case of Art + Com (Figure 1.9); just 40 years ago, a photorealistic globe was established to which users could freely navigate the world, now such capabilities are ubiquitous.

WE HAVE THE TECHNOLOGY, BUT WHAT IS THE DIAGNOSIS?

As Bill Rankin has discussed in the development of cartographic maps, a base map is,

53

> *simultaneously a model for the institutional organization of cartography and the social organization of geographic research: it implies a hierarchy of mapmaking agencies and a linear progression from surveying to the base map to higher-order analysis.*
>
> (Rankin, 2016, p. 26)

Beyond the 2D map, a similar linear progression is implied in constructing UDTs to create 'baseline' information. Yet, the cartographic map has a rich history of geographic representation and political implications based on such description. A UDT is intended as a single source spatial replica connected to the real world (Jones et al., 2020): a digital representation, at a set fidelity, of a physical element, including its behaviour, which is connected and integrated for efficiency (Chapter 2). The World Economic Forum and China Academy of Information and Communication Technology states that a UDT has four typical features:

- **accurate mapping,**
- **analytical insight,**
- **virtual real interaction, and**
- **intelligent intervention,**

for the pursuit of three main future city visions:

- **intensive and efficient urban production and operation,**
- **liveable and convenient urban living spaces, and**
- **sustainable urban ecological environment (WEF, 2022, p. 5).**

Such a schematic of UDTs is indicative of a fluid system of systems approach, which aims to address multiple city aspects such as logistics, energy, construction, and communications, amongst others, through the unification of "surveying and mapping technology, building information modelling (BIM) technology, IoT, 5G, collaborative computing, blockchain, and simulation" (Deng et al., 2022, p. 127; CAICT, 2021).

Masoumi et al. (2023) has utilised a maturity-level model and conducted a meta-literature review and three sub-methods to analyse the state-of-the-art of UDTs. The findings of the study chart an 'emergent field' in DTs, which have yet to fully include 'real-time' data (Li et al., 2020), cyber-physical 'two-way' interactions, and 'autonomous operations' (Saeed et al., 2022). Such a development hierarchy, architecture, and roadmap for digital twins has been established in the UK by the Centre for Digital Built Britain and its vision and Policy for Integration Architecture, Reference Data Libraries, and Foundation Data Model (FDM) for a National Digital Twin (NDT) and Information Management Framework (IMF) between 2018 and 2022 (Heaton & Parlikad, 2020; CDBB, 2022).

UDTs' suppositions of efficiency and reliability need to evolve arguably through the design of the system, the system's analytical capabilities, and system integration in governance. These aspects rely upon frameworks, roadmaps, pilots, and experiments for future possibilities (Petrova-Antonova & Ilieva, 2021). If the current state of the art of UDTs is 'emergent', then progression to high-level attributes relies on forecasting methods based on incremental development and modes of speculation on the trajectory based on current findings and cases (Tao & Qi, 2019). Indeed, much DT literature engages in futuring where these systems are heading through questionable quantified tables and 'attainments'. This is not to say that these forecasts are not always rigorous, and many are grounded upon current approaches and pilot experiments (van der Valk et al., 2021). However, the construction of these socio-technical futures should interest readers in how this knowledge production is created and the pervasive futures being presented alongside adaptive behaviours, organisations, and abilities. Indeed, UDTs require a specific and unique identity due to their cyber-physical twinning and multidisciplinary input in evolving these systems. For example, a New York DT will have very different demands than the DT for Guangzhou, China. Michael Batty (2018) states one of the challenges of digital twins, DT's require specificity and individuality while also being constructed from international and national standards.

> **all physical systems can have a digital equivalent which converges and merges with the system in question. In this sense, a true digital twin running in real-time is no different from the system itself.**
> (Batty, 2018, p. 818)

Batty suggests the digital twin model needs to decouple to simulate new designs and predictions. Modes of speculation can potentially enrich the design of UDTs through prototypes to support predictions, strengthen visions, and articulate new possibilities and foresight. However, digital twins as an immature systems approach are not a panacea for addressing the urgent need for climate-resilient development as described by the UN SDGs and IPCC in the introduction of this chapter. Indeed, as the book charts, there are questions raised around the geographic areas, knowledge formed of those adopting UDT systems, and the cities and territories excluded.

The UN System Framework for Action on Equality, 2016, states that inequalities vary as a term and can be specifically focused on a particular indicator or be a mixture of inequalities such as defined by the U.N. Chief Executives Board for Coordination,

> **income, wealth, wages and social protection, as well as social and legal inequalities where different groups are discriminated against, excluded or otherwise denied full equality. Inequality can also refer to inequality within a country as well as inequality between different countries.**
> (CEB, 2016)

For the OECD, various economic indicators are present across 36 countries as a measure of inequality. For example, 'Regional Innovation' includes regional data on patent and co-patent by technology (fractional count, by inventor and priority year); R&D expenses and R&D employees; and labour force and student enrolment by International Standard Classification of Education (ISCED) level. For UDTs, there are explicit inequalities in the development of information technologies (Pick & Azari, 2008). Inequalities are discussed in this book in relation to ICT investment across World Bank Indicators, alongside the focus on the exclusion of cities that feature in the IMD Smart Cities Index (Global Power City Index, 2022). UDTs is the term used throughout the course of the book, not city digital twins, which preference globally competitive cities, not smart city digital twins, which have undergone extensive critique due to the inequalities they enable, nor connected digital twins, as the urban science is emergent, not established (Angelidou, 2017). UDTs is the term most appropriate as they reflect a more horizontal coverage or a wide variety of geographies and forms of urbanisation. They also have global applications in reflecting broader SDG goals and the technological processes for experimentation using a living lab approach.

Reports, Sensing, and Toolkits

Our City Plans: An incremental and participatory toolbox for urban planning	https://unhabitat.org/our-city-plans-an-incremental-and-participatory-toolbox-for-urban-planning
The Earth Observations Toolkit for Sustainable Cities and Human Settlements	https://eotoolkit.unhabitat.org/
The Intergovernmental Panel on Climate Change	https://www.ipcc.ch/
CEB	https://unsceb.org/
OECD	https://www.oecd-ilibrary.org/
World Bank Development Indicators	https://databank.worldbank.org/source/world-development-indicators
European Network of Living Labs	https://enoll.org/about-us/what-are-living-labs/
INSPIRE	https://knowledge-base.inspire.ec.europa.eu/overview_en

TRL OF DIGITAL TWINS AND GIS – THE MATURITY SPACE

While numerous papers chart the origin of digital twins as a term for product lifecycle management (Glaessgen & Stargel 2012), the majority of existing digital twin systems are akin to the NASA Earth Systems Digital Twin, which are constructed on three main drivers: (1) digital replica, (2) forecasting, and (3) impact assessment.

1. The digital replica is fed by continuous and targeted diverse observations, powered by data assimilation and fusion, and provides an accurate representation of the current state of the system.
2. Forecasting is facilitated by advanced computational capabilities, machine learning (ML) and surrogate modelling, and provides real-time or near-real-time prediction of future states of the system.
3. Finally, impact assessment uses the digital replica and forecasting capabilities with ML, causality, uncertainty quantification and advanced computation, and visualisation capabilities for running large amounts of simulated predictions quickly and at various spatial and temporal scales, opening the door to investigating "what-if" scenarios (NASA, 2022, p. 10).

One of the existing problems for UDTs is the need for a suitable vehicle for the maturity levels of the technology and realistic assessments of the science. NASA developed technology readiness levels (TRL), which are a type of measurement system used to assess the maturity level of a particular technology, which would naturally align to the maturity models currently being sought for UDTs. Each technology project is evaluated against the parameters for each technology level and is then assigned a TRL rating based on the project's progress. There are nine technology readiness levels. TRL 1 is the lowest, TRL 9 is the highest, and the TRLs have been adapted from John C. Mankins (2009) for UDTs here (Figure 1.10). The TRL applied here

CHAPTER 1
DEFINING THE LIVING LAB

Technological Readiness Level>

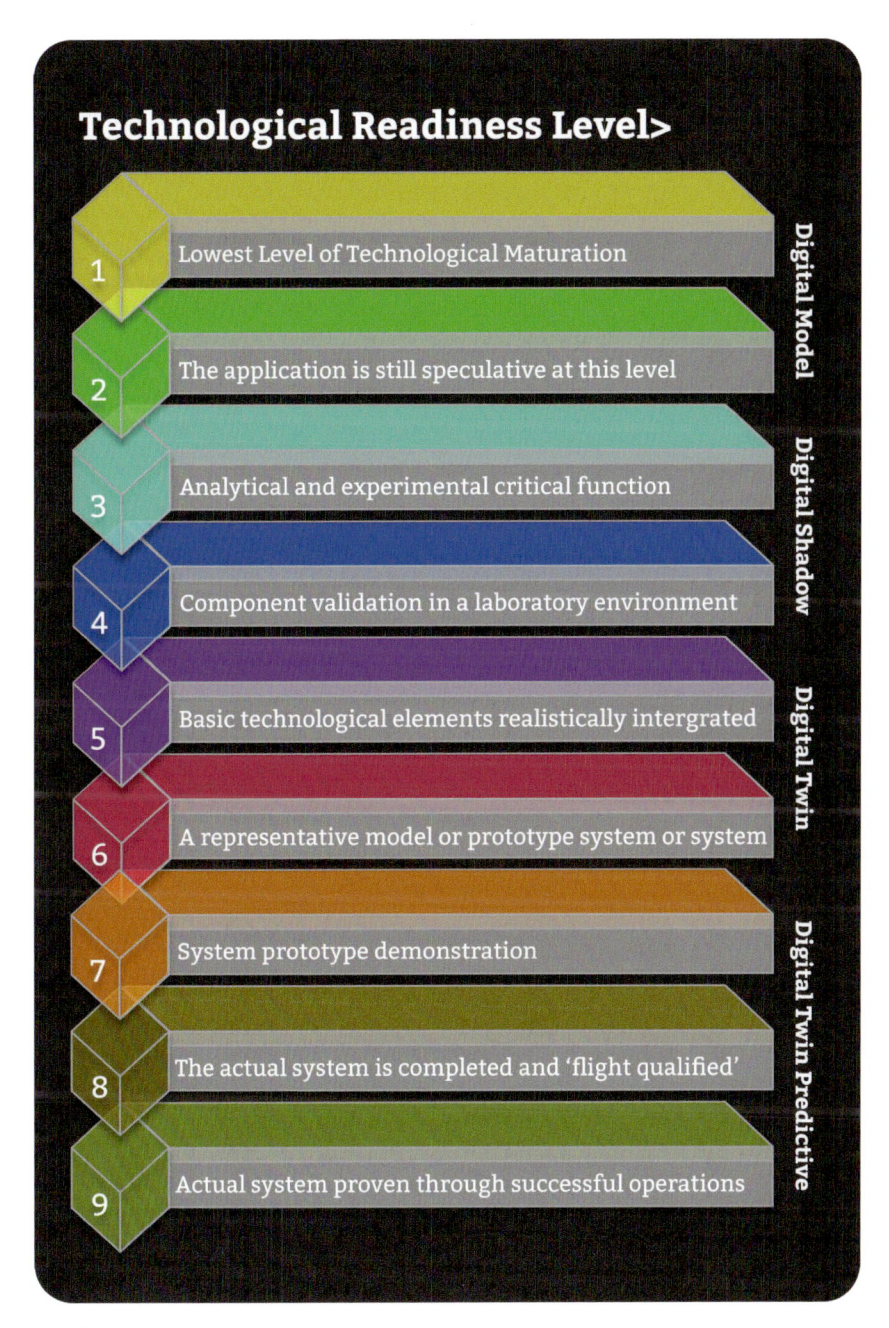

Level	Description	Category
1	Lowest Level of Technological Maturation	Digital Model
2	The application is still speculative at this level	Digital Model
3	Analytical and experimental critical function	Digital Shadow
4	Component validation in a laboratory environment	Digital Shadow
5	Basic technological elements realistically intergrated	Digital Twin
6	A representative model or prototype system or system	Digital Twin
7	System prototype demonstration	Digital Twin Predictive
8	The actual system is completed and 'flight qualified'	Digital Twin Predictive
9	Actual system proven through successful operations	Digital Twin Predictive

Figure 1.10
Author, technological readiness level (TRL) for Digital Twins, 2024.

to UDTs is an interpretation across a variety of sources, including BIM aspects as part of the UK's National Building Standards BIM report (10th ed). The notion of 3D level of detail (LOD) is defined by Filip Biljecki et al. (2014) and the Open Geospatial Consortium, CityGML data standard for 3D urban objects. This chapter will also demonstrate an applied example of design fiction methods in order to speculate at necessary TRL levels and this framing.

TRL 1 – *Lowest Level of Technological Maturation* – Preliminary UDT Frameworks of existing processes, methods and standards such as GIS, Internet of Things (IoT) and Building Information Modelling (BIM). This level would consist of standalone datasets, limited points of interest and baseline geographic coverage, 2D CAD drafts and limited collaboration across public and private sectors.

TRL 2 – *The application is still speculative at this level*: there is no experimental proof or detailed analysis to support the conjecture. At this level, the system could constitute Cloud-based GIS, one-way urban analytics, LOD 1–2 Buildings. This level of TRL is relevant to the Design Fiction method in this chapter, in which speculative prototypes can set a roadmap for further development.

TRL 3 – *Analytical and experimental critical function* and/or characteristic proof-of-concept. Applied examples of this TRL could be city-level household energy atlas or integrated land use plans across combined councils or authorities. Design Fiction is relevant here again as a method for speculative prototyping and world-building for UDTs.

TRL 4 – *Component and/or breadboard validation in a laboratory environment* – connecting various elements (Fuzzy Logic, which is the mathematical representation of vagueness). This could be a demonstration of the relationship between national state mapping datasets and data fusion from citizen-sensed data, which would require various degrees of validation of a number of values for connection (Figure 1.14).

TRL 5 – *The basic technological elements must be integrated with reasonably realistic supporting elements to bring the total applications together.* This TRL would largely be independent in this case; this level would preliminarily integrate land use data, energy models of buildings, and citizen-sensed data hosted on cloud platforms, with 3D visualisations and data dashboards at scale for urban areas and regions. This level would require various servers for processing and a service interface for reporting mechanisms. This level may have data dashboards for governance.

TRL 6 – *A representative model or prototype system or system* – which would go well beyond ad hoc, 'patch-cord' or discrete component level breadboarding – would be tested in a relevant environment. A collection of models and systems set in a geographic context would be interrelated and tested to form a coherent whole, functioning UDT system. The system would not be permanent and would be organised to suit the needs of both organisations and places to which the system 'twins'. Various scenario capabilities are enabled in the UDT for prototype predictions of environmental aspects.

TRL 7 – *System prototype demonstration in a space environment.* At this level, land use, energy use, and citizen-sensed models will be embedded as urban management tools within a council, authority, or regional body. This twinning would inform decision-making and have legal and ethical repercussions regarding the quality and viability of data, monitoring and models. Scenario capability is developed and demonstrated across a range of indictors.

TRL 8 – *The actual system is completed and "flight qualified" through test and demonstration (ground or space).* Training and skills development across a workforce with the UDT is completed, as well as the validation and use of the UDT system, would address system implementation. Validated data and scenarios would inform policy tests and data-driven decisions within the council or authority at a strategic level in committing resources.

TRL 9 – *Actual system "flight proven" through successful mission operations.* A UDT embedded within a council's governance structure informs policy agendas. A UDT roadmap and structure is established and fit for replication in other urban areas globally.

The TRL described indicates the various research problems and development areas of establishing a digital twin. Jan-Frederik Uhlenkamp et al. (2022) have also offered a maturity model for digital twins and a maturity tool to assess capability across a range of DTs, including smart cities, based on seven categories of context, data, computing capabilities, model, integration, control, human-machine interface with 31 ranked characteristics to consider optimal modelling choices and design. Masahiko Haraguchi, Tomomi Funahashi, and Filip Biljecki have also presented a maturity model called CITY-STEPS consisting of eight steps, which have the potential to transfer to rural and non-urban areas and make more socially inclusive city digital twins and is the most developed maturity model in the UDT literature to date (Haraguchi et al., 2024).

While many of the challenges of UDTs are related to design, data, models, and validation, there are also significant human factors to consider regarding the utilisation of UDTs, frameworks for development, and data-driven decision-making.

Data Sources

National Building Standards BIM Report	https://www.thenbs.com/knowledge/national-bim-report-2020
OGC, CityGML	https://www.ogc.org/standard/citygml/
DT Maturity Model	https://dt-maturity.eu/

MODELS AND MODELS AND MODELS...

The possibility is that anything can be 'twinned,' but the fundamental modelling choices and how a twin is operationalised are critically important (Gruen et al., 2019). Given the fluctuation of various terms for competing processes for UDTs and the specificity of the term for readers entering the DT space, it would take substantial work to analyse the team requirements for implementation, the drivers and the organisational management. A model is different from a digital twin (Wright & Davidson, 2020). For UDTs to work, there is a need for not just multidisciplinary collaboration but also transdisciplinary work to address the complexity of the systems and the design of models. Organisational management is an essential aspect of UDT space, but surprisingly, under-discussed. We need more models and technological maturity of UDTs, but the way we work together and work with models needs to happen at the same time. UDTs are not about heterogeneous technological-driven implementation, technological disruption, nor about set methods; UDTs are not value-free scientific experiments (Lei et al., 2023, p. 1). Ramy Al-Sehrawy, Bimal Kumar, and Richard Watson have described four philosophical assumptions underlying the implementation of DTs, which are framed as tech-driven, disruptive, cognitive, and humanistic. The assumptions could contain a mixture of each value set. Humanistic DTs, for example, consist of participatory design, voluntary geographic information, geographically active geo-citizens, and a consciousness of the natural environment. As part of the 'human in the loop' aspect of UDTs, there need to be dual investigations in a range of models to construct federated urban digital twins. As Asaf Tzachor states,

> **Model-based investigations of socio-ecological and socio-technical complexities must go hand-in-hand with critiques informed by the social sciences regarding the blind spots and silences that are embedded in the generalizations of modelling.**
> (Tzachor et al., 2022, p. 825)

Figure 1.11
3D geoinformation group and 3DGI, 3DBAG (TU Delft). © 3DBAG by tudelft3d and 3DG.

The range of models available and applied to digital twins is highly varied. Other models can validate models. If a federated digital twin is to work, it also needs validating models. For example, 3Dbag (Figure 1.11), a national dataset of building modelling covering the whole Netherlands, is used as a baseline for various analyses and modelling, including simulating wind flow, noise pollution, and building retrofitting (Peters et al., 2022). 3Dbag is also incorporated in 3D Basisvoorziening, which includes a national terrain point-cloud model updated yearly (Ávila Eça de Matos, 2023).

In Figure 1.12, QUANT has run three versions and models how workers choose their residence and employment based on the attractiveness of places and the travel costs from their workplaces. The model is constructed from UK Census data from 2011 and journeys to work. The model calibrates its predictions for validation (Batty & Milton, 2021). In longitudinal models (Figure 1.13), researchers from the Living with Machines digital history project (Alan Turing Institute and British Library) developed MapReader, which analyses British Ordnance Survey maps from 1888 to 1913 using computer vision and ML alongside human-trained 'patches' of the maps in order to analyse the impact of railways as an infrastructural technology compared to today (Hosseini et al., 2022; Ahnert & Demertzi, 2023).

CHAPTER 1
DEFINING THE LIVING LAB

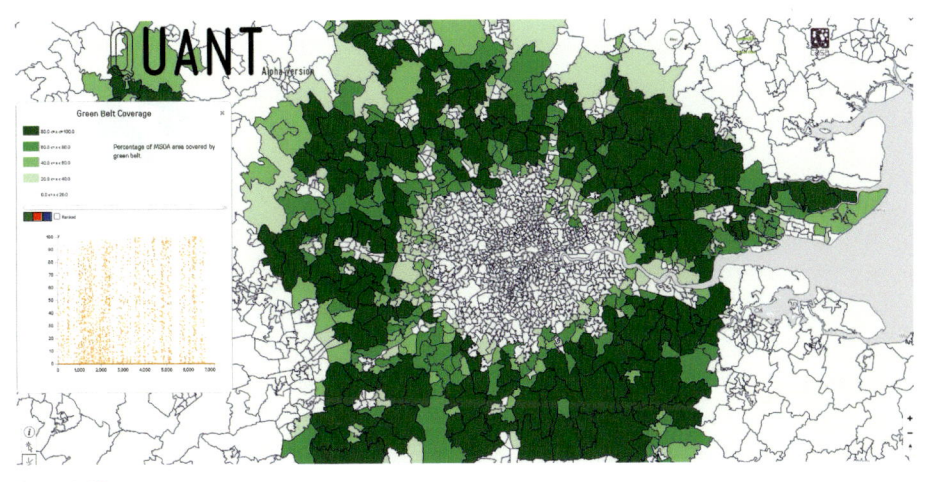

Figure 1.12
Michael Batty and Richard Milton, Quantitative Urban ANalyTics – QUANT, 2020–2021. QUANT is a prototype land use transportation model simulating the location of employment, population, and transport interactions. It is designed to predict the impact of changes to and on these activities. The model predicts the impact of changes in the pattern of urban development as measured by employment, population, transport, and physical land use constraints.

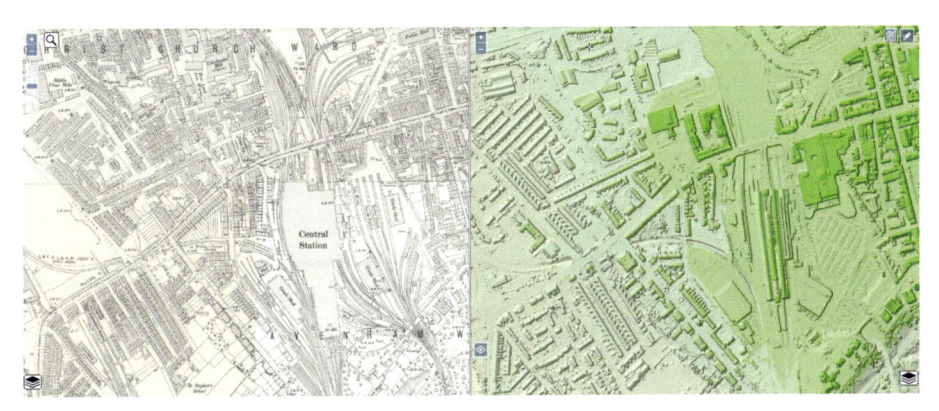

Figure 1.13
Living with Machines, 2017–2023. Map Reader Researchers experimented with the Six-inch to the Mile (1888–1914 maps) of England, Scotland, and Wales to look for railway infrastructure and buildings 'rail space.' 'Reproduced with the permission from the National Library of Scotland.' https://livingwithmachines.ac.uk/the-living-with-machines-report/.

There are numerous smart cities projects across the ASEAN Smart Cities Network comprised of ten member states. Singapore is one such member and has a Smart Nation Initiative. A number of projects operate under this umbrella, relying on various models, such

65

Figure 1.14
SLA, Singapore Land Authority. OneMap is a comprehensive web directory of data and services.

as integrated canopy modelling, for building energy simulations (Ignatius et al., 2019) or the use of datasets for geographic artificial intelligence (GeoAI) for human-centric DTs for the prediction of outdoor comfort (Liu et al., 2023) or the creation of models for urban trees from processing LiDAR, satellite images and quantifying amongst other activities (Gobeawan et al., 2018). In this case, the UDT of Singapore is the most advanced UDT to date and has a collection of federated models and datasets with various analytical possibilities, with the majority merged in a centralised public-facing service (Figure 1.14). The exemplar cases shown here demonstrate a range of modelling choices that could collectively form a UDT, and the next section speculates on other approaches to data collection and modelling through a bottom-up approach (Kopponen et al., 2022).

Urban Digital Twins	
Singapore Land Authority (SLA)	https://www.onemap.gov.sg/
Digital Victoria	https://vic.digitaltwin.terria.io/
3Dbag	https://3dbag.nl/
RailSpace	https://maps.nls.uk/projects/mapreader/#zoom=8.7&lat=54.21504&lon=-3.01246
QUANT	https://www.turing.ac.uk/research/research-projects/quantitative-urban-analytics-quant

URBAN DIGITAL TWINS AND EMBRACING SPECULATION

How can policy goals be evaluated and filtered down into applied cases at the local level, followed by feedback? Reversing the position, what would a bottom-up approach consist of for a UDT? What would the approaches yield in terms of models and predictive qualities? Futurists have utilised scenario and backcasting approaches for futures projection. Project Göteborg 2050 (Gothenburg, Sweden) by Aumna Phdungsilp utilised backcasting as a futuring method for sustainable regional planning in energy, food, urban design, and transportation (2011), and backcasting is an established method in the field of future studies (Dreborg, 1996). For design research, design fiction, coined by Bruce Sterling (2005), has developed through various making and provocation in product design termed 'diegetic prototypes' fully functional technologies in a fictional world (Kirby, 2010; Khan & Zhao, 2021) across IoT, industrial design and film media but has not been applied to UDTs to date.

Design fiction is a mode of futuring through object-orientated ontologies (Harman, 2016) to speculate alternative scenarios and systems. Conflating design, scientific fact, and science fiction (Bleecker, 2009, p. 6), design fiction as an approach has what Meskus and Tikka call constitutive relations in engaging science and technology and socio-material worlds (2022). For Dunne and Raby, speculation focuses on "laws, ethics, political systems, social beliefs, values, fears, and hopes, and how these can be translated into material expressions" (Dunne & Raby 2013, p. 70).

Design fiction processes share modes of speculation apparent in corporate visions and what Coulton et al. term 'vapour-worlds' (things yet to come) (2017) of integrated technology in future everyday life. Design fiction seeks critical intent in constructing new possibilities and scaffolding for alternatives over masking future potentials and market capture (Coulton et al., 2016). Design fiction has three main components:

- **Storytelling – Futuring akin to sci-fiction but a narrative process which embeds diegetic prototypes.**
- **Diegetic prototypes – A fictitious future world created around the prototypes – functional technologies that do not exist, which makes them a matter of concern.**
- **World-building – Future possible worlds constructed in which interdependent and independent objects and environments are created.**

Fundamentally, design fiction is different from scenario-based approaches, which often base projections from the present, and different from science fiction, as the DF method seeks to create a dialogue and social interactions around artefacts and future worlds rather than world-building and narrative for literature. Socio-technical systems thinking emerged in the 1950s from the UK Coal Industry for agile responses to change and technological advances (Trist & Bamforth, 1951). As William Pasmore et al. state, "the evolution of social systems is not keeping pace with the exponential advance of technology, let alone anticipating more pervasive changes yet to come" (William Pasmore et al., 2019, p. 71). Design fiction and science and technology studies (STS) speculative research have real validity in developing UDT future systems, given international visions and speculative hypothesis,

> *A hypothesis is speculative when it aims to be 'productive': its function is to provision epistemic goods through opening new research, or scaffolding the development of theories or experiments, or generating possibility proofs, or providing epistemic links to further knowledge.*
> (Currie, 2023, p. 612)

This chapter argues that a fundamental research gap exists for UDTs, which is an opportunity for design fiction 'world-building,' that is, in

CHAPTER 1
DEFINING THE LIVING LAB

Urban Digital Twin Attributes

Attributes (A)

1) Accurate mapping
2) Analytical insight
3) Virtual real interaction
4) Intelligent intervention

Design Fiction & Research Gap

Research Gap (G)

Narrative (G)

Diegetic Prototypes (G)

World Building (G)

Urban Digital Twin Visions

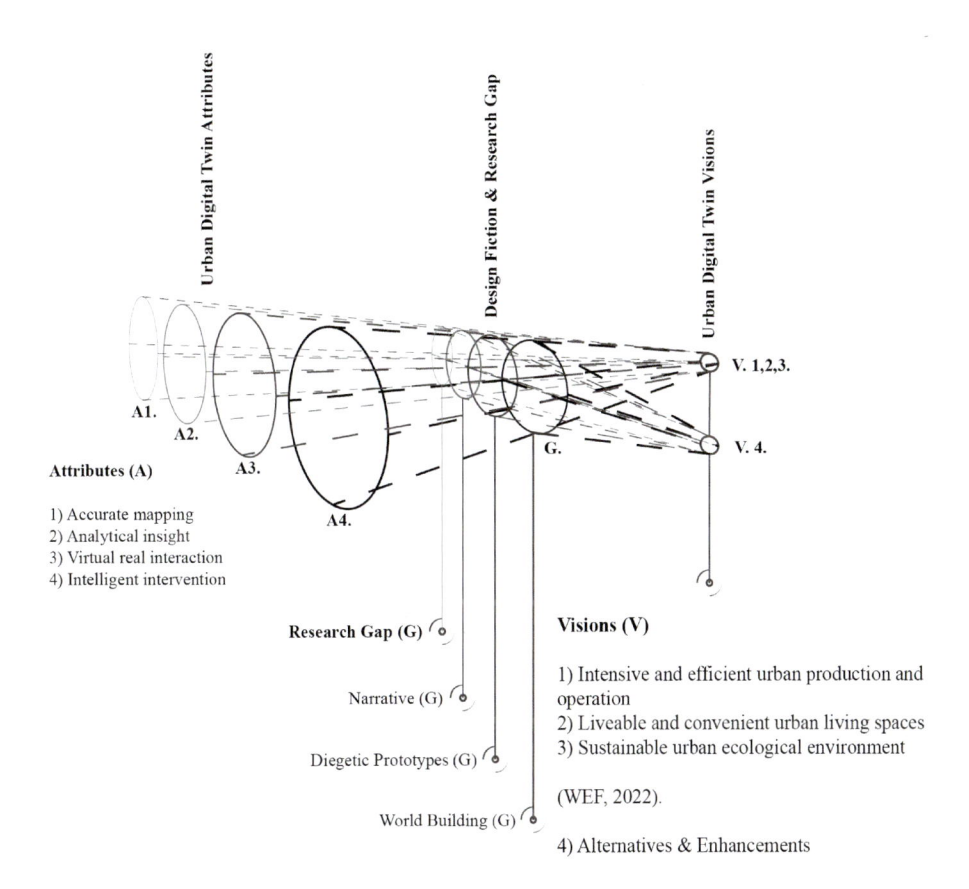

Visions (V)

1) Intensive and efficient urban production and operation
2) Liveable and convenient urban living spaces
3) Sustainable urban ecological environment

(WEF, 2022).

4) Alternatives & Enhancements

Figure 1.15
Authour, 2023. Urban Digital Twins, Design Fiction, and Research Gaps.

terms of placing "importance on the cohesion of the world and how things and people within that world interact" (Coulton et al., 2017, p. 15), which could help address the gap (*G*) between state-of-the-art DT attributes (*A*) and visions (*V*), development stage, efficiencies, and capabilities of DTs (Figure 1.15). Design fiction can potentially scaffold grand city challenges, such as Indonesia's 100 smart cities initiative in 2017 in partnership with government offices of Finland (DIT & Arup, 2022). Design fiction as an approach melds narrative and storytelling alongside world-building, with varying emphasis on foresight (Zaidi, 2019). For example, Alex McDowell, as a production designer, constructed the Minority Report world in the film prior to the story, including the HCI interface artefact of a haptic control room for predictive policing (McDowell, 2019, p. 107) (Figure 1.16).

69

Figure 1.16
Minority Report, directed by Steven Spielberg. Based on the 1956 short story by Philip K. Dick. Seen here, Tom Cruise (as Chief John Anderton, back to camera). He reviews PreCog visions on a monitor. 2002. Screen capture. A Paramount Picture. Getty Images.

World-building, as an intrinsic part of design fiction, aims to create a discursive space for exploration and enables foresight through diegetic prototypes applied to a UDT space, which would, for example, allow the identification of the advances required to achieve the visions and attributes of DTs for urban climate resilience as well as *alternatives* and or enhancements to these visions. Design fiction is not intended to modify the predictive capability of UDTs but to repurpose the modelling systems approach through interactions with stakeholders. In that case, the modeller (urban designer, data science, climate modeller) has a uniquely central role in which speculative hypothesis and *scaffolding* can generate specific advances in proofs and identities for individual primary urban areas and specific twining approaches. Critiques of ICT-centric urbanisation state that economic factors drive smart cities to focus on communication technologies such as 5G, with the limited achievement of environmental SDGs (Boon Lim et al., 2021). Design fiction could, for example, develop novel twinning approaches to coastal and river flooding mitigation to inform climate strategies through open community simulations and scenarios. Alternatively, design fiction through world-building may open larger-scale directions for smart cities such as exploring the more-than-human. As Tan Yigitcanlar et al. (2019) state, "building post-anthropo-centric cities for more-than-human futures might be the last resort for humankind to evolve and avoid extinction in the not-too-distant future" (p. 150). World-building creates a hierarchical process for decision-making through the act of futuring. As part of the science of digital twins, there is a need for research on the design *For* digital twins, explicitly focusing on foresight studies and design fiction for the socio-technical interface of UDT complex systems, forms of simulation and prediction produced, interactions with such modelled futures generated, and alternative worlds.

WORLD-BUILDING

A fundamental modelling challenge and research gap must be overcome through world-building DF to create a diegetic prototype, concentrating on the WEF attribute 1 accurate mapping (simulation) and its feasibility. First, the types of systems modelled (Petrova-Antonova & Ilieva, 2021), and second, the 'real-timeness' of the model data. As Li et al. state, "the continuous flow of real-time city data requires that traditional urban simulation models be renovated towards more accurate, cross-scale and in-time predictions" (2020, p. 315). World-building approaches could help frame the types of modelling architecture and data collected to map the implications of specific dataset uses. For example, in a study of the state of the art of GeoAI, the three most common sectors of application came from remotely sensed images, street view imageries, and transport networks (Wang et al., 2024, p. 4) (Figure 1.17). For example, the UDT could be explored whether it included LSMs, citizen data, such as real-time mobile GPS signals and agent-based modelling (ABM) (Crooks et al., 2021). Such modelling architecture would potentially have significant political implications, such as model output showing significant ecosystem loss or privacy issues from governments

GEODESIGN, URBAN DIGITAL TWINS AND FUTURES

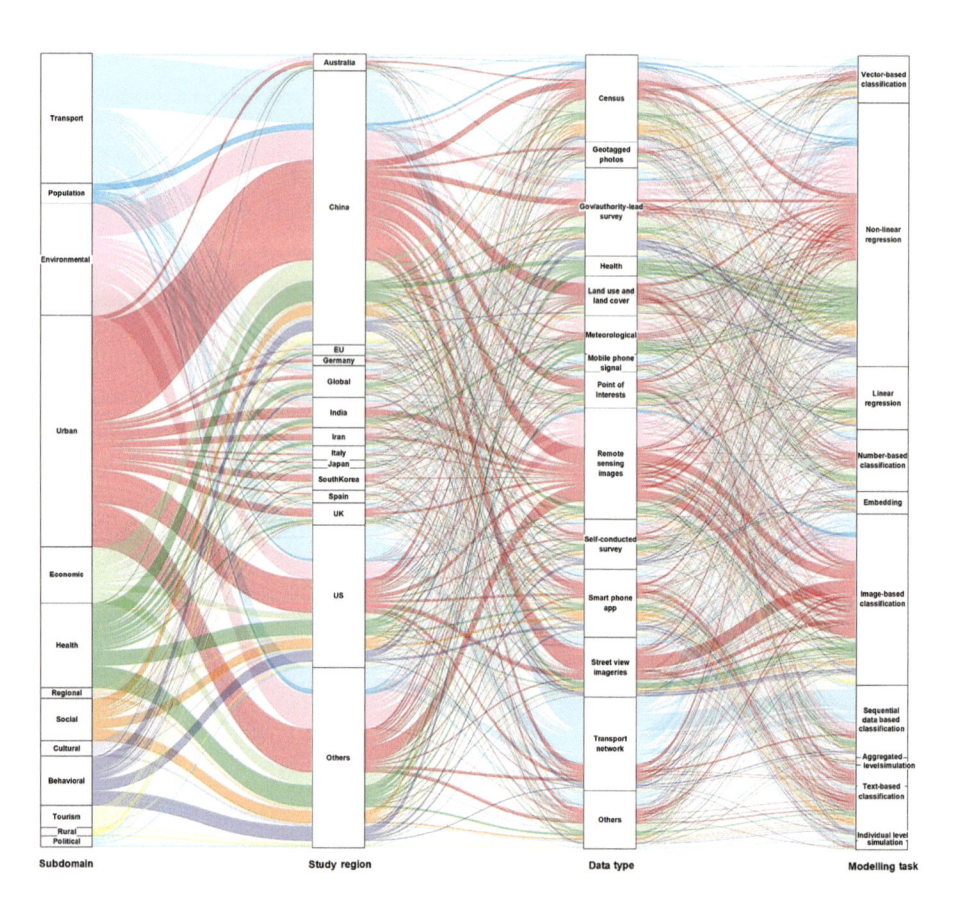

harvesting large-scale citizen data. To construct attribute 1, UDTs, teams, and modellers would generally need to create an inventory of existing sensors and datasets, have extensive fieldwork capability, and access various instruments for geospatial acquisition. To achieve attribute 1, research teams commonly utilise high-resolution satellite data, aerial imagery and LiDAR, and terrestrial surveying equipment alongside integration and assistance from government scale mapping agencies. The pace of technological change and urbanisation for smart-enabled initiatives and urban areas that do not have the resources or technological capacity could radically increase inequalities. As Tzachor et al. state,

Figure 1.17
Siqin Wang et al. GeoAI Sankey Diagram, from 'Mapping the landscape and roadmap of geospatial artificial intelligence (GeoAI) in quantitative human geography: An extensive systematic review,' 2024. This diagram shows the range of applied studies across GeoAI in human geography, which have high degrees of possibility in modelling complexity (Wang et al., 2024).

CHAPTER 1
DEFINING THE LIVING LAB

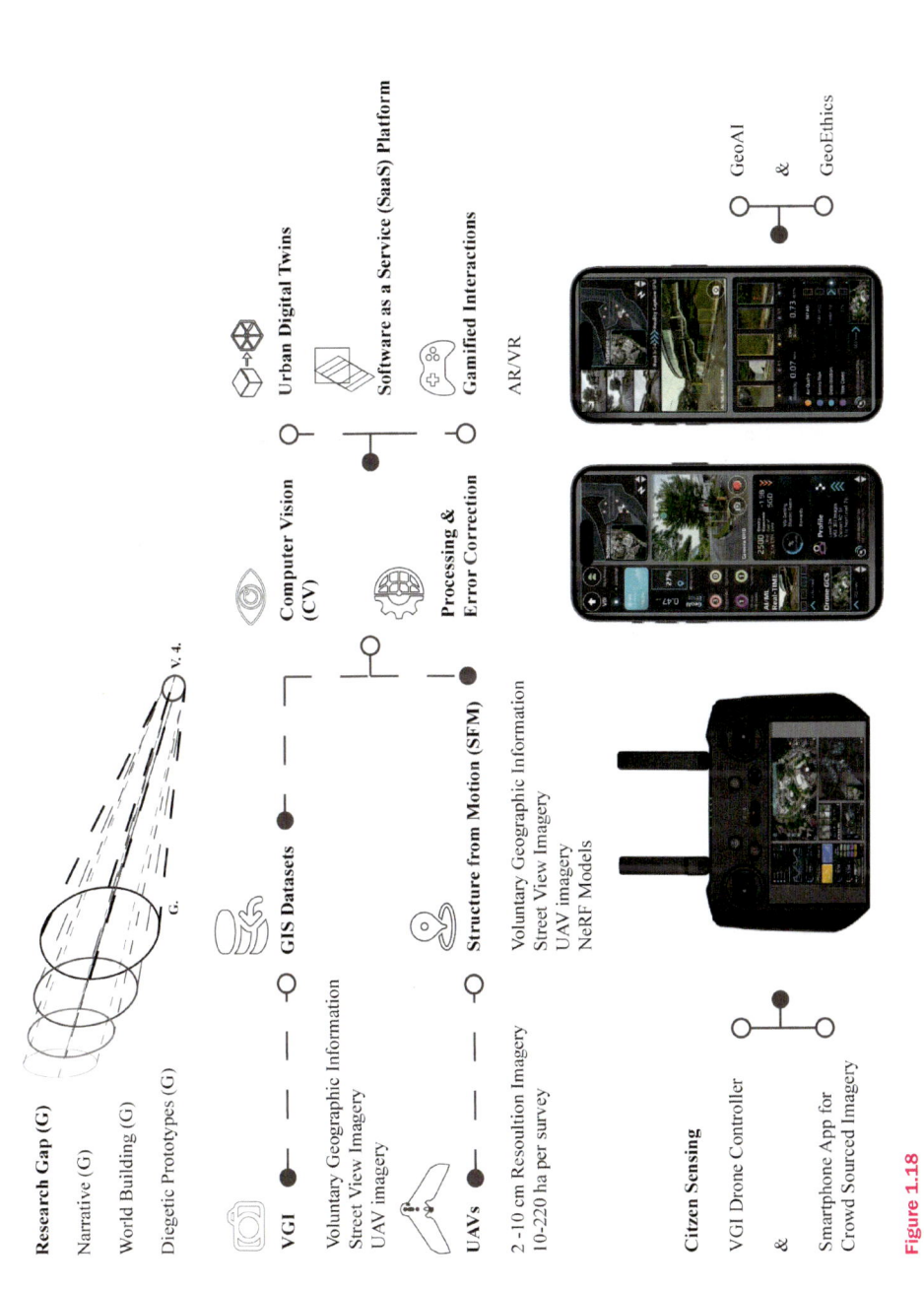

Research Gap (G)
Narrative (G)
World Building (G)
Diegetic Prototypes (G)

VGI
Voluntary Geographic Information
Street View Imagery
UAV imagery

UAVs
2 -10 cm Resoultion Imagery
10-220 ha per survey

Citzen Sensing
VGI Drone Controller
&
Smartphone App for
Crowd Sourced Imagery

GIS Datasets

Structure from Motion (SFM)
Voluntary Geographic Information
Street View Imagery
UAV imagery
NeRF Models

Computer Vision (CV)

Processing & Error Correction

Urban Digital Twins

Software as a Service (SaaS) Platform

Gamified Interactions

AR/VR

GeoAI
&
GeoEthics

Figure 1.18
Authour, 2023. Bottom-up approaches to National Mapping and Citizen Sensed Data.

73

GEODESIGN, URBAN DIGITAL TWINS AND FUTURES

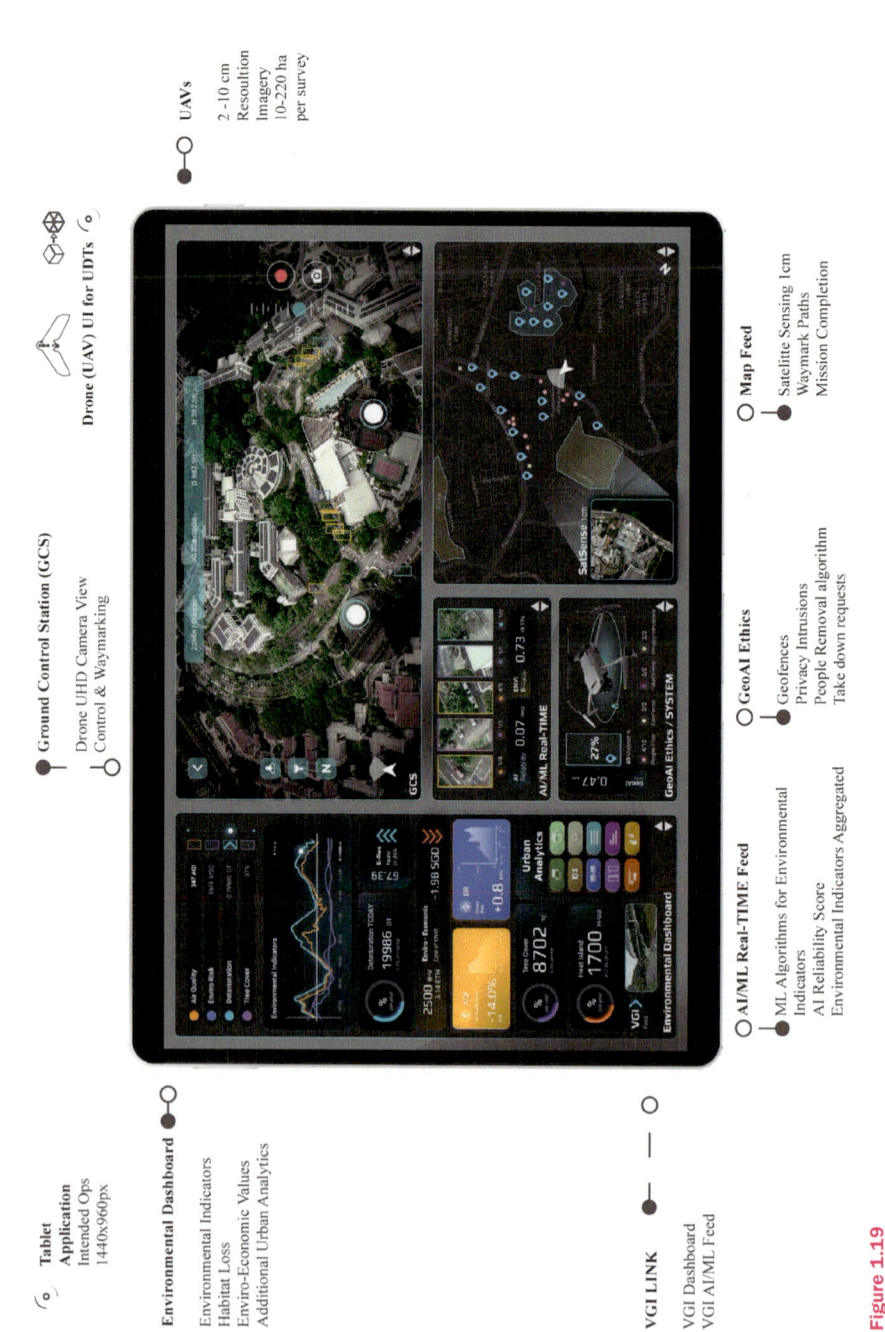

UAVs
2 -10 cm Resoultion Imagery 10-220 ha per survey

Drone (UAV) UI for UDTs

Ground Control Station (GCS)
Drone UHD Camera View
Control & Waymarking

Map Feed
Satellite Sensing 1cm
Waymark Paths
Mission Completion

GeoAI Ethics
Geofences
Privacy Intrusions
People Removal algorithm
Take down requests

AI/ML Real-TIME Feed
ML Algorithms for Environmental Indicators
AI Reliability Score
Environmental Indicators Aggregated

Tablet Application
Intended Ops
1440x960px

Environmental Dashboard
Environmental Indicators
Habitat Loss
Enviro-Economic Values
Additional Urban Analytics

VGI LINK
VGI Dashboard
VGI AI/ML Feed

Figure 1.19
Authour, 2023, Diegetic Prototype, Drone & Smartphone based VGI for a UDT.

CHAPTER 1
DEFINING THE LIVING LAB

Smartphone Application
Intended Ops
390x844px

VGI Dashboard
Privacy Terms & Disclaimer
VGI Contributions
% Mapped
GeoAI Ethics

Map Feed
Satellite Sensing 1cm
Waymark Paths
Mission Completion

Camera UHD
ML Algorithm
People Removal
Stabilisation
Camera Reel
Structure from Motion (SFM)

Enviro Indicators
Economic Loss Value
VGI Public Guide
Rewards System

VGI Gamification
Game Flow for VGI

GPS Positioning

Map Feed
Satellite Sensing 1cm
Waymark Paths

VGI for UDTs

AI/ML Real-TIME Feed
AI/ML Real-TIME Feed
ML Algorithms for Environmental Indicators
AI Reliability Score
Environmental Indicators Aggregated

Environmental Dashboard
Environmental Indicators
Habitat Loss
Enviro-Economic Values
Additional Urban Analytics

Figure 1.19 (Continued)

> ***Adoption of advanced information systems has been led by institutions in high-income economies, which have more resources to invest, skilled labour to mobilize and refined information and communications technology (ICT) infrastructures.***
> (Tzachor et al., 2022, p. 824)

A design fiction approach could create diegetic prototypes of scaled bottom-up mapping approaches from citizen-sensed data for use in GeoAI (Figure 1.18). For example, a diegetic prototype for the purposes of this chapter for a UDT would combine remote imagery using two strategies and one interaction mode. The first strategy would utilise repeat fixed-wing and quadcopter drone surveys (RTK UAV) for 3D mapping (average cover 10–220 ha per survey dependent on model), which could deliver 4D (3D and time-based) imagery focusing on built and natural heritage areas of land to generate indicators of spatial change. The second strategy would use VGI for citizens' smartphones and the use of computer vision (CV) techniques to classify imagery and objects, such as the technique applied by Ito and Biljecki (2021), utilising street view imagery and CV to assess bike-ability of urban blocks through the establishment of a bike-ability index (Figure 1.19). Both UAV-derived imagery and VGI and street-level imagery have low costs and near-global coverage and can be established as open data by attribution to researchers CC-BY. There are data quality and management issues in using VGI, but the potential across a range of areas is substantial, including disaster management, and continual updates may offer the frequency necessary for a near 'real-time' update of ground conditions (Haworth & Bruce, 2015). In the process of construction, a noted feature would need to be developed around GeoAI ethics and the real-time ML techniques, meaning people-blurring algorithms, take down requests from the public, and other screening tools. Following the fusion and modelling of the two data types, purposed for an urban planning perspective, the interaction, from the 'bottom-up' datasets, would be through a process of 'gamification.' Gamification is game design elements applied to non-game contexts (Deterding et al., 2011).

Gaming methodologies could generate future scenarios, interactions, and participatory practices already evidenced by the proliferation of cloud-based GIS, urban planning serious games (Khan & Zhao, 2021; Hudson-Smith & Shaker, 2022). The outline schematic prototype, which includes a gamified interaction for user profiles, scaffolds alternative approaches to UDT creation through low-cost citizen-generated data and could be an intrinsic element of a larger system. For example, this citizen-sensed data would be incorporated in larger cloud-based UDTs mentioned here through progression of the TRLs. However, this schematic relies on modellers overcoming technical challenges in the diegetic prototype in collection methods and using these 'bottom-up' approaches, addressing data reliability, constructing models alongside other open-source datasets, and designing a system for interaction via serious games (Chapter 5).

LIMITATIONS

This first advance in world-building for UDTs would require much more refinement of scenarios, geography, and needs via stakeholder input, interaction, and advancement into a demonstratable system via GIS and games to demonstrate the validity and examination of WEF attribute 1. Indeed, further development of the UDT DF diegetic prototype is necessary to fully articulate the viability of simulating urban conditions via design fiction, address various visions articulated in UDT research, and consider alternative or enhancement possibilities that UDTs could enable towards SDG goals. Design fiction would not seek to invalidate the reliability of various predictions from UDTs and urban analytics but offer a criticality to change the directionality and intentionality of the systems via modellers. The world-building activity would also be necessary for design fiction's other three WEF attributes to consider the entirety of DT systems to scaffold fuller alternatives and new critical intentions. Given the novel combination of design fiction and UDT, there is a lack of precedent of methods, though the potential of socio-technical interactions and knowledge formation has huge potential. However, as many UDTs are emergent at various stages of capability across a range of geographies, there is a danger of increasing digital inequalities through UDT creation primarily for tier 1 cities, global cities rather than a variety of urban places (Joss et al., 2019, Ferré-Bigorra et al., 2022). However, due to the embryonic nature of UDTs, there is timeliness for experimentation for a range of geographical UDTs, given that such critical approaches may yield important alternatives of the utility of UDTs, which can address pressing urban issues before resources are fully committed at later TRL stages.

TOWARDS A CITIZEN SCIENCE – WE CAN ONLY DO THIS TOGETHER

As UDTs become more mature, a systematic framework for a future UDT utilising design fiction methods could unlock alternative perspectives from various disciplines for DT creation and citizen-generated data. Platforms such as Mapillary (Figure 1.20) and Open Street Map (OSM) have been established since 2006 and 2013. There is a range of studies that evidence of the capacity of crowd-sourced imagery applied across a wider range of urban analyses which can contribute to a UDT (Figure 1.20) (Brewster, 2015; Shen et al., 2021; Zeng et al., 2023, Zheng & Amemiya, 2023). The case study for Virtual Gothenburg Lab on social sustainability in development by László Sall Vesselényi and Sanna Klefbom demonstrates an augmented reality (AR) prototype for citizen dialogue (Figure 1.21). However, VGI only provides donated imagery and not the variety of sensing necessary to contribute to a large federated UDT. Technological challenges and research gaps can be understood through world-building and diegetic prototypes to inform current real-world UDT roadmaps and visions as well as offer new criticalities and alternatives. Such speculative methods enable explicit details on the state-of-the-art UDT research (Ketzler et al., 2020) as evidence to create feasible diegetic prototypes, open alternative directions from

CHAPTER 1
DEFINING THE LIVING LAB

Figure 1.20
Mapillary, NeRF of Cities. Eagles' Nests Landscape Park, Poland and Swiss National Museum, Zurich, Switzerland.

the predicative qualities of the systems, and open new intentions of urban modelling even for municipal areas without extensive smart city investment through 'bottom-up' data collection methods. Such directionality is critical in addressing inequalities in urban areas without UDT predictive systems. Through the process of world-building UDTs, there is significant potential for critical insights and alternatives for designing the future to address pressing climate challenges and global policy challenges.

Sources

Citizen Science	https://eu-citizen.science/
Mapillary	https://www.mapillary.com/
Piero	https://giro3d.org/

79

CASE STUDY: RISE RESEARCH INSTITUTES OF SWEDEN, VIRTUAL GOTHENBURG LAB, COORDINATED BY VISUAL ARENA

Social Sustainability in Urban Development was a pilot project led by RISE Research Institutes of Sweden within the project Virtual Gothenburg Lab, coordinated by Visual Arena. The pilot explored how the urban digital twin Virtual Gothenburg could support citizen dialogue and participation in urban development processes in new and engaging ways.

The outcome was a prototype of an augmented reality (AR) app as an interface to Virtual Gothenburg for inhabitants and city officials. The intention was to support community place-making for citizens and help them raise a stronger common voice about their values and activities, as well as for city officials to communicate with citizens and better understand social and cultural values of a place. With AR, the digital twin could be visualised as a virtual layer in physical places.

For the inhabitants, this app allows contributing information to the urban digital twin by virtually placing predefined 3D models of objects in the environment and sharing experiences about a place in multimedia formats (see Figure 1.21). Inhabitants are encouraged to interact with information shared by others by modifying or adding objects and suggestions already placed in the virtual environment. In addition to citizen-generated content, citizens would be able to access official city data through the app, such as statistics and proposals generated by city officials.

As for the city officials, the app would provide place-specific communication with citizens in and about urban development processes and support existing dialogue and participatory processes. City officials would be able to publish time- and place-specific requests,

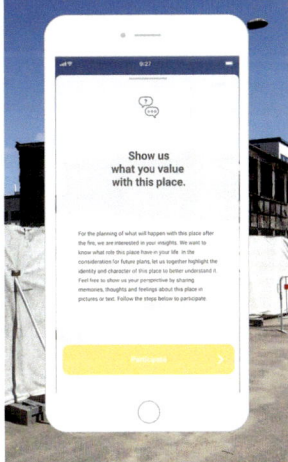

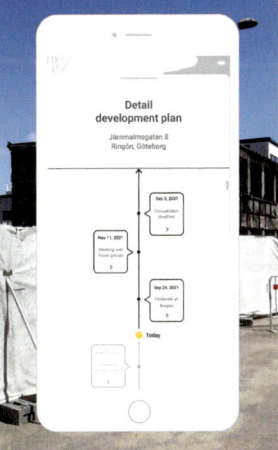

Figure 1.21
László Sall Vesselényi and Sanna klefbom, 2023.

inquiries, and questions to citizens connected to specific urban planning processes. Additionally, city officials would have access to and be able to respond to content published by citizens.

The pilot was inspired by an ambition to challenge technocratic conceptions of urban digital twins with a more participatory, bottom-up approach. Citizens could be understood as a city's most important

81

'sensors,' each contributing individual and collective perspectives on what the city is. That is, they can provide the richest insights on the city and its developments, even when the insights are wildly contradictory. The combination of the personal interface and the virtual character of citizen input attempts to allow these contradictory experiences and realities of what the city of Gothenburg is to coexist and enrich one another and be as messy as a real city.

https://visualarena.lindholmen.se/en/project/virtual-gothenburg-lab

REFERENCES

Ahnert, R., & Demertzi, L. (2023). *Living with machines final report*. The Alan Turing Institute. (Accessed 30/07/2023) https://livingwithmachines.ac.uk/the-living-with-machines-report/.

Allan, M., Rajabifard, A., & Foliente, G. (2024). Urban regeneration and placemaking: A Digital Twin enhanced performance-based framework for Melbourne's Greenline Project? *Australian Planner*, 1–11. https://doi.org/10.1080/07293682.2024.2342793.

Angelidou, M. (2017). Smart city planning and development shortcomings. *TeMA Journal of Land Use, Mobility and Environment*, 10. (Accessed 12/04/2018) http://www.serena.unina.it/index.php/tema/article/view/4032.

Ávila Eça de Matos, B. (2023). *Digital twins for cities*. Technische Universiteit Eindhoven.

Barzegar, M., Rajabifard, A., Kalantari, M., & Atazadeh, B. (2021). A framework for spatial analysis in 3D urban land administration – A case study for Victoria, Australia. *Land Use Policy*, 111, 105766. https://doi.org/10.1016/j.landusepol.2021.105766.

Batty, M. (2018). Digital twins. *Environment and Planning B: Urban Analytics and City Science*, 45(5), 817–820.

Batty, M., & Milton, R. (2021). A new framework for very large-scale urban modelling. *Urban Studies*, 58(15), 3071–3094. https://doi.org/10.1177/0042098020982252.

Biljecki, F., Ledoux, H., Stoter, J., & Zhao, J. (2014). Formalisation of the level of detail in 3D city modelling. *Computers, Environment and Urban Systems*, 48, 1–15. ISSN 0198-9715. https://doi.org/10.1016/j.compenvurbsys.20.

Bleecker, J. (2009). Design fiction A short essay on design, science, fact and fiction. (Accessed 20/09/2022) https://systemsorienteddesign.net/wp-content/uploads/2011/01/DesignFiction_WebEdition.pdf.

Bloch, T., & Sacks, R. (2018). Comparing machine learning and rule-based inferencing for semantic enrichment of BIM models. *Automation in Construction*, 91, https://doi.org/10.1016/j.autcon.2018.03.018.

Boon Lim, S., Abdul Malek, J., Yusof Hussain, M., Zurinah, T., & Saman, N.H. (2021). SDGs, smart urbanisation and politics: Stakeholder partnerships and environmental cases in malaysia. *Journal of Sustainability Science and Management*, 16(4), 190–219.

Brewster, S. (2015). Build a 3-D virtual world with this crowdsourced map of the real world. *MIT Technology Review*.

CAICT. (2021). *Smart city industry mapping research report*. http://www.caict.ac.cn/english/research/whitepapers/202112/t20211228_394675.html.

CDBB. (2018–2022). https://www.cdbb.cam.ac.uk/what-we-did/national-digital-twin-programme.

Coulton, P., Lindley, J., & Akmal, H.A. (2016). Design fiction: Does the search for plausibility lead to deception? In P. Lloyd, & E. Bohemia (Eds.), *Proceedings of design research society conference 2016* (pp. 369–384). (Proceedings of DRS 2016, vol. 1). Design Research Society.

Coulton, P., Lindley, J.G., Sturdee, M., & Stead, M. (2017). Design fiction as world-building. In *Proceedings of research through design conference 2017*. https://doi.org/10.6084/m9.figshare.4746964.

Crooks, A., Heppenstall, A., Malleson, N., & Manley, E. (2021). Agent-based modeling and the city: A gallery of applications. In W. Shi, M.F. Goodchild, M. Batty, M.P. Kwan, & A. Zhang (Eds.), *Urban informatics. The urban book series*. Springer. https://doi.org/10.1007/978-981-15-8983-6_46.

Currie, A. (2023). Science & speculation. *Erkenn*, 88, 597–619. https://doi.org/10.1007/s10670-020-00370-w.

Deng, Z., Chen, Y., & Yang, J. et al. (2022). Archetype identification and urban building energy modeling for city-scale buildings based on GIS datasets. *Building Simulation*, 15, 1547–1559. https://doi.org/10.1007/s12273-021-0878-4.

Department of International Trade, & ARUP. (2022). *Smart city handbook Indonesia*. https://www.events.great.gov.uk/ereg/newreg.php?eventid=200246885&.

Deterding, S., Dixon, D., Khaled, R., & Nacke, L. (2011). From game design elements to gamefulness: Defining "gamification". In *Proceedings of the 15th International Academic MindTrek Conference: Envisioning Future Media Environments (MindTrek '11)* (pp. 9–15). Association for Computing Machinery, New York, NY, USA. https://doi.org/10.1145/2181037.2181040

Dreborg, K.H. (1996). Essence of backcasting. *Futures*, 28(9), 813–828, ISSN 0016-3287. https://doi.org/10.1016/S0016-3287(96)00044-4.

Dunne, A., & Raby, F. (2013). *Speculative everything: Design, fiction, and social dreaming*. MIT Press.

Ferré-Bigorra, J., Casals, M., & Gangolells, M. (2022). The adoption of urban digital twins. *Cities*, 131, 103905, ISSN 0264-2751. https://doi.org/10.1016/j.cities.2022.103905.

Fisher, R.A., & Koven, C.D. (2020). Perspectives on the future of land surface models and the challenges of representing complex terrestrial systems. *Journal of Advances in Modeling Earth Systems*, 12, e2018MS001453. https://doi.org/10.1029/2018MS001453.

Glaessgen, E., & Stargel, D. (2012). The digital twin paradigm for future NASA and U.S. air force vehicles. *53rd AIAA/ASME/ASCE/AHS/ASC* structures, structural dynamics and materials conference; 20th AIAA/ ASME/AHS adaptive structures conference; 14th AIAA (pp. 1–14). American Institute of Aeronautics and Astronautics, Reston, VA.

Global Power City Index. (2022). https://mori-mfoundation.or.jp/english/ius2/gpci2/index.shtml.

Gobeawan, L., Lin, E.S., Tandon, A., Yee, A.T.K., Khoo, V.H.S., Teo, S.N., Yi, S., Lim, C.W., Wong, S.T., Wise, D.J., Cheng, P., Liew, S.C., Huang, X., Li, Q.H., Teo, L.S., Fekete, G.S., & Poto, M.T. (2018). Modeling trees for virtual Singapore: From data acquisition to citygml models. *International Archives of the Photogrammetry, Remote Sensing and Spatial Information Sciences*, XLII-4/W10, 55–62, https://doi.org/10.5194/isprs-archives-XLII-4-W10-55-2018.

Gruen, A., Schubiger, S., Qin, R., Schrotter, G., Xiong, B., Li, J., Ling, X., Xiao, C., Yao, S., & Nuesch, F. (2019). Semantically enriched high resolution lod 3 building model generation. *International Archives of the Photogrammetry, Remote Sensing and Spatial Information Sciences*, XLII-4/W15, 11–18. https://doi.org/10.5194/isprs-archives-XLII-4-W15-11-2019, 2019.

Haraguchi, M., Funahashi, T., & Biljecki, F. (2024). Assessing governance implications of city digital twin technology: A maturity model approach. *Technological Forecasting and Social Change*, 204, 123409. https://doi.org/10.1016/j.techfore.2024.123409.

Harman, G. (2016). *Object-orientated onotlogy: A new theory of everything*. Penguin.

Haworth, B., & Bruce, E. (2015). A review of volunteered geographic information for disaster management. *Geography Compass*, 9, 237–250. https://doi.org/10.1111/gec3.12213.

Heaton, J., & Parlikad, A.K. (2020). Asset information model to support the adoption of a digital twin: West Cambridge case study. *IFAC-PapersOnLine*, 53(3), 366–371, ISSN 2405-8963. https://doi.org/10.1016/j.ifacol.2020.11.059.

Hosseini, K., Wilson, D.C.S., Beelen, K., & McDonough, K. (2022). MapReader: A computer vision pipeline for the semantic exploration of maps at scale. In *Proceedings of the 6th ACM SIGSPATIAL international workshop on geospatial humanities (GeoHumanities '22)* (pp. 8–19). Association for Computing Machinery, New York, NY. https://doi.org/10.1145/3557919.3565812.

Hudson-Smith, A., & Shakeri, M. (2022). The Future's Not What It Used To Be: Urban Wormholes, Simulation, Participation, and Planning in the Metaverse. Urban Planning, 7(2), 214-217. https://doi.org/10.17645/up.v7i2.5893

Ignatius, M., Wong, N.H., Martin, M., & S Chen, S. (2019). Virtual Singapore integration with energy simulation and canopy modelling for climate assessment. *IOP Conference Series: Earth and Environmental Science*, 294, 012018.

IPCC. (2023). In H. Lee, & J. Romero (Eds.), *Climate change 2023: Synthesis report*. Contribution of Working Groups I, II and III to the Sixth Assessment Report of the Intergovernmental Panel on Climate Change (pp. 35–115). IPCC, Geneva, Switzerland. https://doi.org/10.59327/IPCC/AR6-9789291691647.

Irajifar, L. (2022). Digital solutions for resilient cities: A critical assessment of resilience in smart city initiatives in Melbourne, Victoria. In A. Sharifi, & P. Salehi (Eds.), *Resilient smart cities*. The Urban Book Series. Springer, Cham. https://doi.org/10.1007/978-3-030-95037-8_11.

Ito, K., & Biljecki, F. (2021). Assessing bikeability with street view imagery and computer vision. *Transportation Research Part C*, 132, 103371.

Jones, D., Snider, C., Nassehi, A., Yon, J., & Hicks, B. (2020). Characterising the digital twin: A systematic literature review. *CIRP Journal of Manufacturing Science and Technology*, 29, Part A, 36–52, ISSN 1755–5817. https://doi.org/10.1016/j.cirpj.2020.02.002.

Joss, S., Sengers, F., Schraven, D., Caprotti, F., & Dayot, Y. (2019). The smart city as global discourse: Storylines and critical junctures across 27 cities. *Journal of Urban Technology*, 26(1), 3–34. https://doi.org/10.1080/10630732.2018.1558387.

Ketzler, B., Naserentin, V., Latino, F., Zangelidis, C., Thuvander, L., & Logg, A. (2020). Digital twins for cities: A state of the art review. *Built Environment*, 46(4), 547–573(27).

Khan, T.A., & Zhao, X. (2021). Perceptions of students for a gamification approach: Cities skylines as a pedagogical tool in urban planning education. In D. Dennehy, A. Griva, N. Pouloudi, Y.K. Dwivedi, I. Pappas, M. Mäntymäki (Eds.), *Responsible AI and analytics for an ethical and inclusive digitized society*. I3E 2021. Lecture notes in computer science, vol. 12896. Springer, Cham. https://doi.org/10.1007/978-3-030-85447-8_64.

Kirby, D. (2010). The future is now: Diegetic prototypes and the role of popular films in generating real-world technological development. *Social Studies of Science*, 40(1), 41–70. https://doi.org/10.1177/0306312709338325.

Kopponen, A., Hahto, A., Kettunen, P., Mikkonen, T., Mäkitalo, N., Nurmi, J. & Rossi, M. (2022). Empowering citizens with digital twins: A blueprint. *IEEE Internet Computing*, 26(5), 7–16. https://doi.org/10.1109/MIC.2022.3159683.

Langenheim, N., Sabri, S., Chen, Y., Kesmanis, A., Felson, A., Mueller, A., Rajabifard, A., & Zhang, Y. Adapting a digital twin to enable real-time water sensitive urban design decision-making. (2022). *International Archives of the Photogrammetry, Remote Sensing and Spatial Information Sciences*, XLVIII-4/W4-2022, 95–100. https://doi.org/10.5194/isprs-archives-XLVIII-4-W4-2022-95-2022.

Lei, B., Janssen, P., Stoter, J., & Biljecki, F. (2023). Challenges of urban digital twins: A systematic review and a Delphi expert survey. *Automation in Construction*, 147, 104716, ISSN 0926-5805. https://doi.org/10.1016/j.autcon.2022.104716.

Li, W., Batty, M., & Goodchild, M.F. (2020). Real-time GIS for smart cities. *International Journal of Geographical Information Science*, 34(2), 311–324. https://doi.org/10.1080/13658816.2019.1673397.

Liu, P., Zhao, T., Luo, J., Lei, B., Frei, M., Miller, C., & Biljecki, F. (2023). Towards human-centric digital twins: Leveraging computer vision and graph models to predict outdoor comfort. *Sustainable Cities and Society*, 93, 104480. https://doi.org/10.1016/j.scs.2023.104480.

Mankins, John C. (2009). Technology readiness assessments: A retrospective. *Acta Astronautica*, 65(9–10), 1216–1223. ISSN 0094-5765. https://doi.org/10.1016/j.actaastro.2009.03.058.

Masoumi, H., Shirowzhan, S., Eskandarpour P., & James Pettit, C. (2023). City digital twins: Their maturity level and differentiation from 3D city models. *Big Earth Data*. https://doi.org/10.1080/20964471.2022.2160156.

McDowell, A. (2019). *Journal of Futures Studies*, 23(3), 105–112. https://doi.org/10.6531/JFS.201903_23(3).0009.

Meskus, M., & Tikka, E. (2022). Speculative approaches in social science and design research: Methodological implications of working in 'the gap' of uncertainty. *Qualitative Research*. https://doi.org/10.1177/14687941221129808.

NASA. (2022). *Earth Systems Digital Twin (ESDT) workshop report* (Co-organized with Earth Science Information Partners, ESIP), October 26–28, 2022 (Virtual).

Peters, R., Dukai, B., Vitalis, S., van Liempt, J., & Stoter, J. (2022). *Photogrammetric Engineering & Remote Sensing*, 88(3), 165–170(6). https://doi.org/10.14358/PERS.21-00032R2.

Petrova-Antonova, D., & Ilieva, S. (2021). Digital twin modeling of smart cities. In T. Ahram, R. Taiar, K. Langlois, & A. Choplin (Eds.), *Human interaction, emerging technologies and future applications III*. IHIET 2020. Advances in Intelligent Systems and Computing, vol. 1253. Springer, Cham. https://doi.org/10.1007/978-3-030-55307-4_58.

Phdungsilp, A. (2011). Futures studies' backcasting method used for strategic sustainable city planning. *Futures*, 43(7), 707–714, ISSN 0016-3287. https://doi.org/10.1016/j.futures.2011.05.012.

Pick, J.B., & Azari, R. (2008). Global digital divide: Influence of socioeconomic, governmental, and accessibility factors on information technology. *Information Technology for Development*, 14(2), 91–115. https://doi.org/10.1002/itdj.20095.

Rae, A., & Wong, C. (Eds.) (2021). *Applied data analysis for urban planning and management.* London Sage Publications.

Rankin, B. (2016). After the Map – Cartography, Navigation, and the Transformation of Territory in the Twentieth Century. University of Chicago Press.

Saeed, Z.O., Mancini, F., Glusac, T., & Izadpanahi, P. (2022). Future city, digital twinning and the urban realm: A systematic literature review. *Buildings*, 12, 685. https://doi.org/10.3390/buildings12050685.

Shen, Y., Xu, Y., & Liu, L. (2021). Crowd-sourced city images: Decoding multidimensional interaction between imagery elements with volunteered photos. *ISPRS International Journal of Geo-Information*, 10(11), 740. https://doi.org/10.3390/ijgi10110740.

Wang, S., Huang, X., Liu, P., Zhang, M., Biljecki, F., Hu, T., Fu, X., Liu, L., Liu, X., Wang, R., Huang, Y., Yan, J., Jiang, J., Chukwu, M., Naghedi, S.R., Hemmati, M., Shao, Y., Jia, N., Xiao, Z., Tian, T., Hu, Y., Yu, L., Yap, W., Macatulad, E., Chen, Z., Cui, Y., Ito, K., Ye, M., Fan, Z., Lei, B., & Bao, S. (2024). Mapping the landscape and roadmap of geospatial artificial intelligence (GeoAI) in quantitative human geography: An extensive systematic review. *International Journal of Applied Earth Observation and Geoinformation*, 128, 103734, ISSN 1569-8432. https://doi.org/10.1016/j.jag.2024.103734.

Sterling, B. (2005). Shaping things (Mediaworks Pamphlets).

Stott, L. (2022). *Partnership and transformation: The promise of multistakeholder collaboration in context* (1st ed.). Routledge. https://doi.org/10.4324/9781003199434.

Tao, F., & Qi, Q. (2019). Make more digital twins. *Nature*, 573, 490–491.

Trist, E., & Bamforth, K. (1951). Some social and psychological consequences of the longwall method of coal-getting. *Human Relations*, 4, 3–38.

Tzachor, A., Sabri, S., & Richards, C.E. et al. (2022). Potential and limitations of digital twins to achieve the Sustainable Development Goals. *Nature Sustainability*, 5, 822–829. https://doi.org/10.1038/s41893-022-00923-7.

Uhlenkamp, J.-F., Hauge, J. B., Broda, E., Lütjen, M., Freitag, M., & Thoben, K.-D. (2022). Digital twins: A maturity model for their classification and evaluation. *IEEE Access*, 10, 69605–69635. https://doi.org/10.1109/ACCESS.2022.3186353.

U.N. (2022). Indicator 11.3.2: Proportion of cities with a direct participation structure of civil society in urban planning and management that operate regularly and democratically. (Accessed 06/01/2023) https://odd-dashboard.cd/en/11-3-2/

U.N. CEB. (2016). Leaving no one behind: Equality and non-discrimination at the heart of sustainable development leaving no one behind: equality and non-discrimination at the heart of sustainable development. (Accessed 10/01/2023) https://unsceb.org/sites/default/files/imported_files/CEB%20equality%20framework-A4-web-rev3_0.pdf

van der Aalst, W.M.P., Hinz, O., & Weinhardt, C. (2021). Resilient digital twins. *Business & Information Systems Engineering*, 63, 615–619. https://doi.org/10.1007/s12599-021-00721-z.

van der Valk, H., Haße, H., Möller, F. et al. (2021). Archetypes of digital twins. *Business & Information Systems Engineering*. https://doi.org/10.1007/s12599-021-00727-7.

WEF. (2022, 20 April). Digital twin cities: Framework and global practices. https://www.weforum.org/reports/digital-twin-cities-framework-and-global-practices/.

William Pasmore, W., Stu Winby, S., Susan Albers, S., Vanasse, M., & R. (2019). Reflections: Sociotechnical systems design and organization change. *Journal of Change Management*, 19(2), 67–85. https://doi.org/10.1080/14697017.2018.1553761.

Wright, L., & Davidson, S. (2020). How to tell the difference between a model and a digital twin. *Advanced Modeling and Simulation in Engineering Sciences*, 7, 13. https://doi.org/10.1186/s40323-020-00147-4.

Yigitcanlar, T., Marcus Foth, M., & Kamruzzaman, M. (2019). Towards post-anthropocentric cities: Reconceptualizing smart cities to evade urban ecocide. *Journal of Urban Technology*, 26(2), 147–152. https://doi.org/10.1080/10630732.2018.1524249.

Zaidi, L. (2019). Worldbuilding in science fiction, foresight and design. *Journal of Futures Studies*, 23(4), 15–26.

Zheng, X., & Amemiya, M. (2023). Method for applying crowdsourced street-level imagery data to evaluate street-level greenness. *ISPRS International Journal of Geo-Information*, 12(3), 108. https://doi.org/10.3390/ijgi12030108.

Zhou, W., Pickett, S.T.A., & McPhearson, T. (2021). Conceptual frameworks facilitate integration for transdisciplinary urban science. *npj Urban Sustainability*, 1, 1. https://doi.org/10.1038/s42949-020-00011-9.

CHAPTER 2

TOWARDS URBAN DIGITAL TWINS

URBAN DIGITAL TWINS OVERVIEW

Urban digital twins (UDTs) are intended as a single source spatial replica connected to the real world and have real potential for analytics, prediction, and management of urban spaces under climate change through a living lab approach and potentially alongside bottom-up approaches, as we have seen in the previous chapter. Digital twins are a digital representation at a set fidelity of a physical element, including its behaviour, which is connected and integrated for efficiency. UDTs are applied to cities and urban environments and are fundamentally part of the development of smart cities. However, digital twins have a range of archetypal meanings and have a particular use case and application, which leads to a diversity of terms (van der Valk et al., 2021). First considered by NASA and coined by Michael Grieves in 2003, digital twins as a subject has risen exponentially over the last ten years, and this development journey has been well mapped (Grieves & Vickers, 2017). The range of digital twin applications is vast, and digital twins are being developed for urban planning and smart city contexts. Digital twins can be situated as the next phase of smart cities and are intended to provide analysis and insights into the complexity of urban environments. Thus, we could create a history using a backcasting method (Chapter 3) (Figure 2.1) of the first iterations of smart cities to more recent digital twin projects at various scales. The arrival of UDTs could be thematically divided into four phases: first, the birth of CAD and GIS; secondly, the emergence of the Internet of Things (IoT), smart cities 1.0, city information models (CIMs), and building information modelling (BIM); thirdly, digital twins for cities, smart cities 2.0; and fourthly, connected digital twins for cities and the built environment.

We could arguably correlate UDTs as part of the continuing development and forms of the smart city. However, Michael Batty states that the smart city is based on the variable technology that defines it, "thus the question 'what and where is the smart city?' not only has

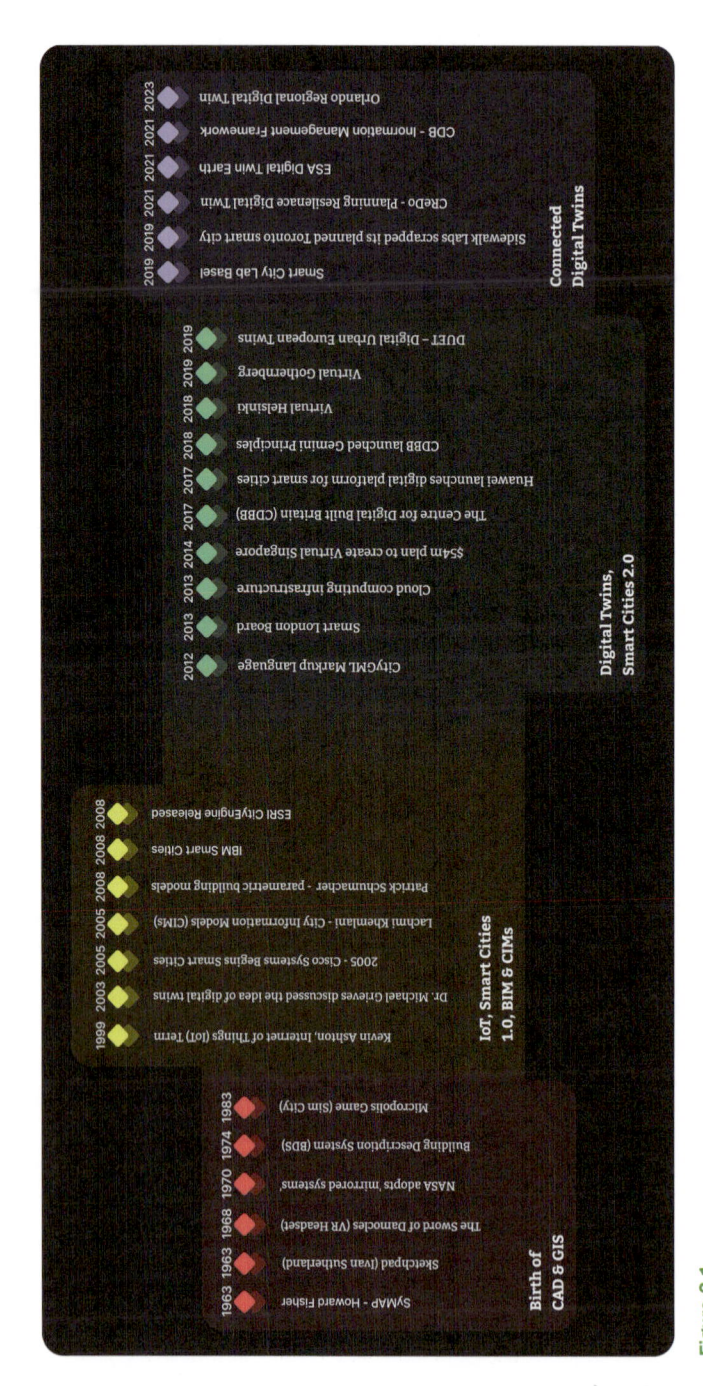

Figure 2.1
Authour, A Short Timeline of Digital Twins, 2024.

no answer, it is also ill-defined, largely because smartness or intelligence is a process, not an artefact or product" (Batty, 2018, p. 178). That said, smart cities are arguably a set of loose terms symbolising ICT processes applied to urban settings. Identifying smart cities can be undertaken through the IMD-SUTD Smart City Index (SCI), which is an annual account of ICT technologies and urban impact and charts the various levels of implementation and global performance of cities with Singapore, Zurich, Oslo, and Taipei performing in the top tier, which also have various urban digital twin projects in development.

Since the advent of the personal computer and software applications developed for real space, architecture, planning, and engineering fields have sought automated and intelligent ways of accessing and generating (big) data to inform design with the implementation of embodied sensors in the urban fabric (Batty, 2018, p. 176). This chapter defines several urban digital twin processes and current state of the art for digital twins applied in urban planning contexts, including applied GIS. Presently, digital twins of urban environments are informed through various fields as part of urban informatics, urban analytics, geographic data science, and geo-computation (making geographical decisions about how best to tackle a real-world problem (Brunsdon & Singleton, 2015). In essence, urban informatics, urban analytics, and geo-computation are terms with overlapping research communities across a variety of disciplines applied in the real world using a variety of approaches and methods for sensing, working with big spatial data, and the modes in which people utilise data, and the communication of findings, models, and predictions. We are in a period of large volumes of ubiquitous 4D urban data (3D and time-based) captured from embedded, connected, and remote sensors. The ability of participants to engage in various forms of immersion and interaction with this data has developed through cloud-based GIS platforms (Souza & Bueno, 2022). Yet, these areas are often unconnected, fragmented, and unequal. The two phases of the history of smart cities have caused technological inequalities, and large territories could benefit from urban digital twin approaches. Equally, big data and artificial intelligence (AI) research are concentrated and dispersed to high-technology areas (Lutz, 2019). Digitisation is advancing urban infrastructure (Broo & Schooling, 2021); technological autonomy in cities and high-performance geographic systems, geospatial AI, machine learning techniques and streamlined procedures, virtual and augmented simulations, and feedback systems have emerged concurrently with climatic challenges, pandemics, and increasing global urbanisation.

Current major digital twins established to date include Virtual Helsinki (case in this chapter), Virtual Singapore, and Virtual Gothenburg, Sweden, with Chalmers University amongst other cities, all focused on architecture, planning, and smart city developments (Hämäläinen, 2021). In addition, many major urban settlements are developing their road maps for UDTs as part of broad digitisation drives from national governments. For example, Munich is supported by a national geoformation strategy for Germany through a LiDAR acquisition programme.

As part of broader efforts for Digital Twinning in Vienna, Austria, a specific project in development from the Survey and Mapping Department is creating the Digital geoTwin for Vienna, focusing on geodata as the basis for these wider twinning efforts. Building on a long history of surveying and mapping from analogue stereo-photogrammetry beginning in 1938 to the present day, contemporary data demands require new processes in CIM creation and ways of aligning datasets. The Digital geoTwin Vienna changes the traditional modelling paradigm in which various inputs are correlated in order to form a 3D model and creates the 3D model first, from which geo-datasets are then derived. This strategy ensures full temporal and contentual coherence between the Digital geoTwin and all derived geo-datasets. Fundamentally, this novel modelling effort creates a base, a CIM, in order for wider DT efforts to be situated.

Figure 2.2
Digital geoTwin Vienna, Austria, 2019.

CIM focuses on modelling geo-objects and information to generate an intelligent data enriched digital replica, an intelligent virtual city which can then be used for simulations, models and analysis.

(Lehner & Dorffner, 2020, p. 67)

The Vienna geoTwin establishes this through rigorous geometric and semantic geodata combinations that constitute the CIM overall by linking the objects of the Digital geoTwin with further data and information, e.g. census data, socio economic data, energy consumption data, maintenance management data, etc. It also raises issues in terms of the level of accuracy of survey data seeking new approaches and model maintenance. The Vienna geoTwin data model is a blueprint for wider CIM creation (Lehner et al., 2024). The broader Digital Twinning efforts in Vienna envision applications in analysing, planning and maintaining the city by constant feedback loops between physical and virtual space, i.e. the digital twin. Digital twins, in comparison, are intended as cyber-physical systems that involve computation, control, and networking. They are embedded in the physical environment as part of a broader movement of automation with the potential to disrupt many sectors and purport to drive efficiency and innovation. This disruption equally applies to the role of applied GIS (Song et al. 2017).

However, a digital twin is a loose term in the urban context, and this book argues for using the term urban digital twin as a more

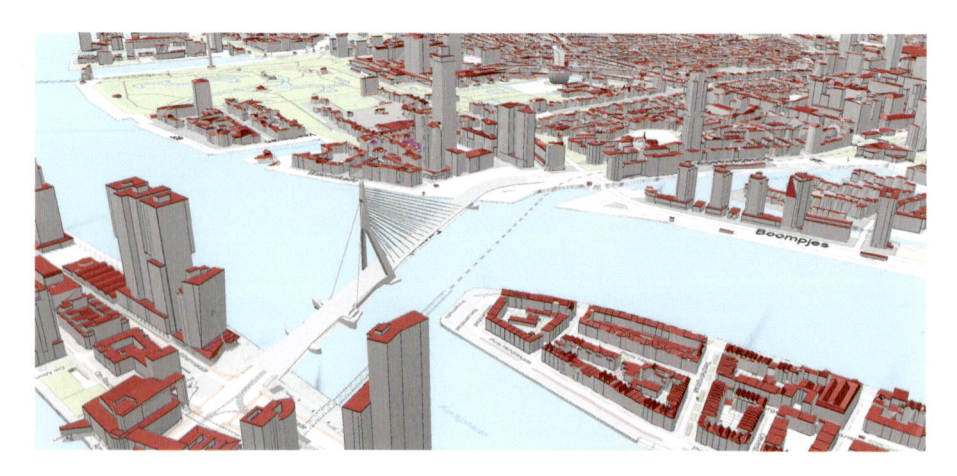

Figure 2.3
Municipality of Rotterdam, Netherlands, 3DRotterdam. Beginning in 2014, 3D Rotterdam is an open-source platform of 3D LOD2 buildings, LiDAR, Tree maps, and rootballs and street furniture following CityGML standards. https://www.3drotterdam.nl/#/.

inclusive approach to address large global urbanisation challenges and not just as a continuation of smart city projects concentrated on global cities. As a concept, an urban digital twin is more than a 3D GIS model but is sometimes mistakenly understood in this way (Ketzler et al., 2020). Commercial marketing of 'city digital twins' is also 'over promise' and provides urban 3D GIS models and isolated and unconnected products. Alternatively, digital twins may be understood as being connected to individual buildings or structures as part of the BIM processes (Zomer et al., 2021). BIM manages information for building construction, often including post-construction maintenance as a total lifecycle. Emergent technologies, systems, and processes such as these must be carefully evaluated and raise various challenges and research issues. Digital twins, from their conceptual origin, suit aerospace and manufacturing arenas more than spatial environments, which have higher uncertainties, complexity, and connectivity requirements (Bettencourt, 2024). Urban environments are inherently messy and complex, making implementing systems and modelling choices more difficult (Qiuchen, 2021).

We need to move beyond 'smart' debates and focus on the socio-technological relationships of UDTs. Margarita Angelidou cites five main criticisms of smart cities in general and some dangers that could apply to the next generation of digital twins. First, their conceptual and methodological ambiguity; second, ICT-centric and corporate-driven utopian visions; third, the overlooking of citizens and stakeholders, splintered urbanism, and unequal representation; fourth, privacy and security concerns; and finally, lack of long-term vision for sustainable urban development adapted to local needs (Angelidou, 2017, p. 79). These five factors are important considerations as to why such resources are invested in digital twin development as embodied IT sensors are required; resources are needed for virtualisation, management and security, and regulatory compliance (Charitonidou, 2022). M&M projected that the global digital twin market will be valued at USD 3.1 billion in 2020 and reach 48.2 billion by 2026 with a common annual growth rate (CAGR) of 58%, with the most extensive digital twin areas being automotive and transportation. The scope of operation and data of digital twins is wide and varied, with many different identities.

For example, discussions and vision settings for alternative twins exist for 'environmental' digital twins (Blair, 2021). Such environmental DTs have a unique role, and the European Space Agency (ESA) Digital Twin of Earth offers some exciting possibilities. This environmental aspect of DTs stems from more extended research areas in satellite remote sensing and geosciences (Klippel et al., 2021). Spatial DTs are essentially mirrored and connected to the environment in which they are based. Thus, when discussing DTs, each has a very different identity and purpose, forming part of a broader national ecosystem. For example, the Centre for Digital Built Britain (CDBB) established the Gemini Principles, a broad set of principles considering the range of sensing, data management, sense-making, and decision-making alongside a pathway to an Information Management Framework that may help inform high-level strategies for future UDTs (CDBB, 2021). The contemporary nature of DTs is why the GIS field is a part of a paradigm shift in geographic computational design, analysis, prediction, and management of future places.

TOWARDS VIRTUALISATION OF COMPLEXITY – BASELINE, FREQUENCY, AND DATA COLLECTION

The current state of the art of UDTs for geospatial and urban planning contexts are, in fact, discussions and developments of CIMs (Omrany et al., 2022).

First termed for Autodesk University in 2005 by Lachmi Khemlani, CIMs are virtual representations of the real world with information layers and various analytical capabilities and simulation precursors to whole UDT systems (Xu et al., 2021, Xue et al., 2021). In the previous chapter, a technological readiness level (TRL) for UDTs demonstrated the progressive steps and also aligns with other researchers' maturity models, such as Masahiko Haraguchi, Tomomi Funahashi, and Filip Biljecki, CITYSTEPS consisting of eight steps, which could provide socially inclusive city digital twins (Haraguchi et al., 2024). CIMs are 3D GIS models of urban environments which combine 3DGIS with BIM (Geo – BIM) (de Laat & van Berlo, 2011). GeoBIM links data in architecture, engineering, and construction (AEC) and connects designed data, construction data, and parametric models to municipalities and planning departments. As Noardo et al. state,

By using GeoBIM, the rich information produced by design and correctly georeferenced into the 3D city model through a tested methodology can be effectively and objectively used in its completeness by the Municipality and the same checked, data can converge into the 3D city model to update it.
(2020, p. 211)

There is potential for updating and digitalising urban planning from one 'single source of truth' through a UDT systems approach. However, there are still significant research challenges for GeoBIM and interoperability in planning (Arroyo Ohori et al., 2018). Readers may wish to consult the GeoBIM benchmark and training datasets for the state of the art. For CIMs, a recent systematic literature review by Letícia Souza and Cristiane Bueno of CIMs revealed digital twins and smart cities ambiguities in the terms (2022). Many CIMs have, as a standard, a connected 3D baseline model for supplementing and layering information. However, 3D baselines are not always necessary. The discussions around 3D baselines and the implementation of CIMs are dominated by several software platforms in GIS, such as ESRI CityEngine and 3D modelling software such as Rhino and Blender alongside games engines such as Unity and Unreal Engine for VR experiences. These modelling aspects are just one part of CIMs, and one aspect of developed UDTs. These baseline models usually contain a mixture of aerial imagery derived from 2D information or generated semantic 3D buildings at a level of detail (LOD), reality meshes, and some environmental assets such as trees and street furniture, transport corridors, and infrastructures such as roads and bridges as well as rural geographies (Qian & Leng, 2021). The authors developed a pilot CIM to explore this potential in 2019 (Cureton & Hartley, 2023).

James H. Clark first proposed LOD for 3D computer graphics in 1976 and later formed part of the CityGML ontology in how the 3D data exactly mirrors the real world (Clark, 1976). For CIMs, information layers are 'connected' data, also known as 'connected threads,' that inform the analytic possibilities based on the base model such as wind flow simulations (García-Sánchez et al., 2021). For example, existing buildings and a proposed building for planning impact and analysis. An example of a connecting thread is the Ordnance Survey, UK, Unique Property Reference Number (UPRN), the identifier used to link buildings to addresses. This dataset is a 'golden' thread; for example, it allows additional layers to be connected, such as the heritage classification and record of a building or energy performance.

Thus, baseline data is a fundamental precursor for the 'virtualisation' and establishment of a CIM or urban digital twin but does not constitute the entirety of the digital twin system, as shown in the Digital Twin Consortium®, which has published indicative schematic of a DT system. As shown, the systems require a wide variety specialism and a requirement for multi-disciplinary areas and collaboration in order to develop a system of systems approach (Figure 2.4). The DT has several subsystems, platforms, services, and interfaces. For example, the internal dynamics of the DT would be structured around (red) areas of security, trust, and governance with some external sources informing policy. The IT/OT platform (green) would host and store data, providing networking and having several application programming interface (APIs) on a software platform such as ESRI ArcGIS online platform and ESRI ArcGIS urban service. The DT would feature a virtual representation (yellow)

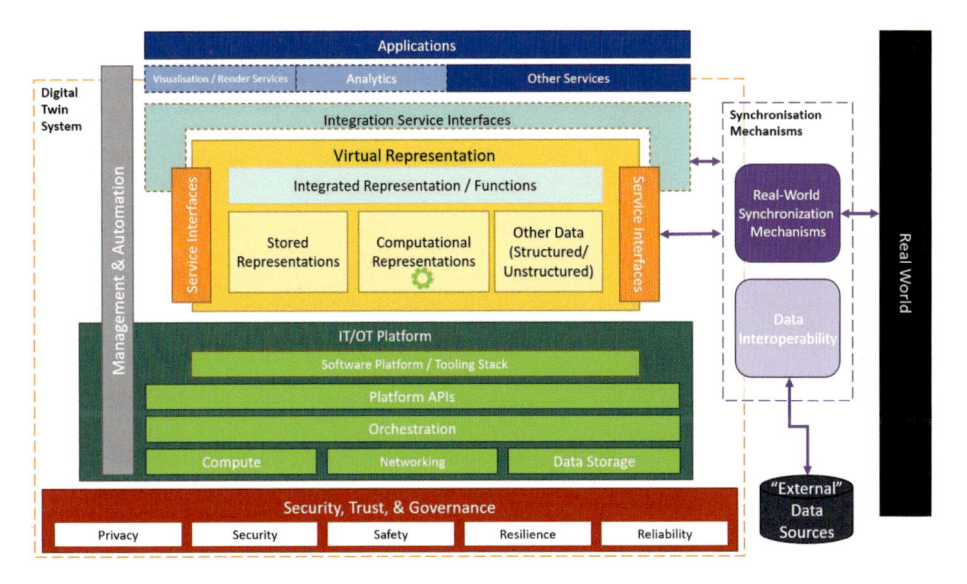

Figure 2.4
**Digital Twin System, Platform Stack Architectural Framework: An Introductory Guide of the Digital Twin Consortium®
(2023, p. 9). Copyright of the Digital Twin Consortium, the Authority in Digital Twin, all rights reserved, 2022.**

synchronised to the real world (black). Services from the virtual representation (blue) include analytics, visualisation, and applications. The schematic indicates the systems of systems approach required for a DT. In addition to emerging frameworks, cross-disciplinary standards, localised needs, and deployments must also be considered (Figures 2.5 and 2.6).

In the case of a system of systems approach within a broad UDT, the global engineering company Mott Macdonald proposes a framework which consists of data collection, including reality capture, IoT and social data, and data management, what is termed a single source of truth hosted locally or cloud-based as a city-data platform (Wildfire, 2020). Furthermore, the third layer of analytics offers modelling and optimisation. Finally, city solutions are proposed for various sectors and optimised to advise enterprises, infrastructure, and citizens. The function of this schematic is commercially intended for services by the company, but the inherent conceptual framework is similar to the outline schematic of a DT system of systems. Digital twins

Figure 2.5
Hartley and Cureton, Lancaster City Information Model (LCIM), 2019–2021.

conceptually are exact replicas, and in an urban context, this means that data gathered must have some real-time or regular feedback to the model(s) if true to the concept. Li et al. suggest four broad

CHAPTER 2
TOWARDS URBAN DIGITAL TWINS

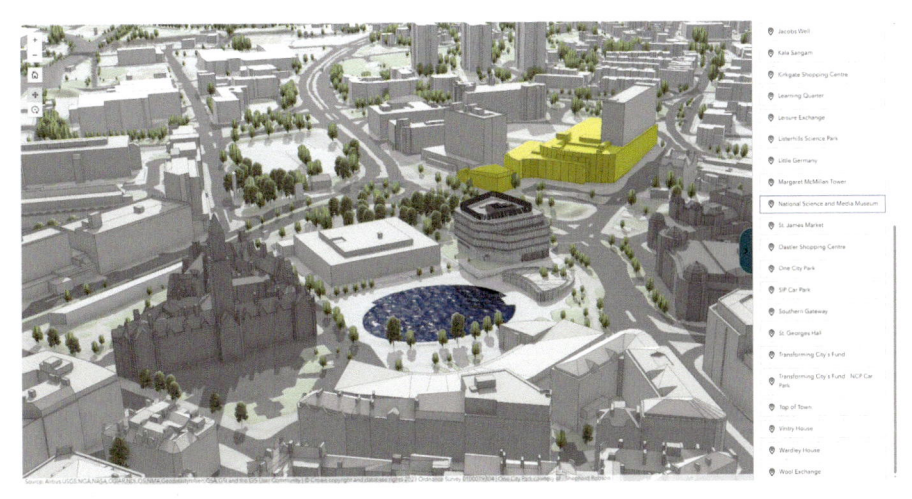

Figure 2.6
Bradford city centre 3D modelling from Virtual Bradford (University of Bradford – Visualising Heritage and City of Bradford Metropolitan District Council) supplemented by LOD1 background detail derived from Ordnance Survey mapping and data, Blue Sky tree data, and One City Park BIM model courtesy of Sheppard Robson. Published in ESRI ArcGIS Online to highlight development opportunities at UKREiiF 2023. Another image shows a schematic showing part of Little Germany Conservation Area captured as part of Virtual Bradford, representing meshed point cloud data captured using drone imagery and mobile mapping through to a textured model. This open digital twin helps to underpin data-driven decision-making for the City of Bradford and represents a strategic partnership between the University of Bradford (Visualising Heritage) and the City of Bradford Metropolitan District Council (Department of Place), made possible with initial seed funding from the EU SCORE programme.

factors for achieving real-time GIS in smart cities: geospatial data, urban simulation, city-data dashboards, and GIS processing (2020). 'Real-timeness' poses several challenges, and here, we can consider one particular aspect of these factors: geospatial data.

Creating UDTs requires careful design of the form and frequency of data collection. This involves the use of a wide variety of acquisition systems and instruments. For example, one popular element of digital twin ambition is the construction of 3D models and meshes of urban environments at a specified LOD. To generate 3D datasets, aerial imaging may be required at specific resolutions. However, a team would need to decide on flight plans, image resolution, and the frequency of repeat flights within their operational budget. This frequency may increase if the settlement is surveyed for significant development projects and infrastructure. In addition, specialist embedded sensors may be required, and other instruments could be used to supplement this information. Therefore, virtualisation could involve a combination of instruments, such as light aircraft, drones, and vehicle-mounted sensors. Results processing would require what is termed 'data fusion': "The fundamental concept of data fusion is the extraction of the best-fit geometry data as well as the most suitable semantic data from existing datasets" (Stankutė & Asche, 2009). In an example from Nottingham City Council, the CIM model is being explored using a variety of survey techniques and fused data, with reality and semantic models captured from mobile LiDAR mapping, oblique aerial imagery, open source, and governmental data amongst many other fields (Figure 2.7).

As a research problem and hypothesis (Figure 2.8), how would you create baseline 3D spatial data from scratch for a region in the UK covering 3,079 km^2 that would involve multiple instruments? In this case, indicative coverage of a single survey is shown: light aircraft (1,150 ha), fixed-wing drones (220 ha), quadcopter drones (100 ha), vehicle-mounted scanners, and terrestrial scanners (8 ha). In addition, material from the public termed volunteered geographic information (VGI) is also shown to supplement the 'real-time' aspect of virtualisation. However, the rate of return would be unknown, as would the quality of the material. Thus, the virtual representation of the environment is one aspect of the urban digital twin but creates several research problems in that a digital twin is intended to have 'real-time' updates and have bi-directional capability, to which extent is limited through the survey design and frequency. In a planning

CHAPTER 2
TOWARDS URBAN DIGITAL TWINS

Figure 2.7
Mick Dunn, Laura Pullen, Geographical Information Services, Nottingham City Council, 2024. Gary Priestnall, University of Nottingham, Projection Augmented Relief Model (PARM) of Nottingham city, UK, part funded by the Ministry of Housing, Communities & Local Government via Nottingham City Council (NCC), and the Regional Innovation Fund. The PARM is being used for analytics for urban heat (led by Simon Gosling and Emily O'Donnell) and carbon storage (led by Sam Booth).

105

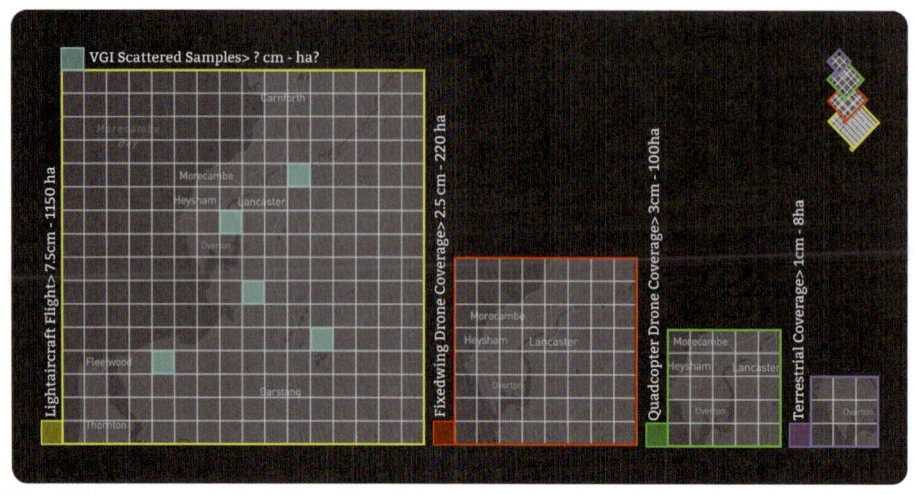

Figure 2.8
Authour, Survey Plan, 2024.

context, this can be mitigated by required digital model submission of major schemes by applicants to planning portals to allow updating, open-access published 'as-built' models or through more economical means of survey such as drones or through the inclusion of public participatory GIS (PPGIS) or VGI methods (see Chapter 4).

For the City of Zurich Digital Twin, the city council sought digitalisation in one programme through a digital twin focus on BIM and urban planning for climate change, including 4D aspects. The primary data set stems from 2014 and has undergone several iterations and encompasses LiDAR and a cadastral survey,

> *The spatial data models represent reality, which determine structure and content independent of a specific spatial data system. The formal descriptions of the characteristics of spatial data are based on metadata, such as origin, content, structure, validity, timeliness, accuracy, usage rights, access options or processing methods.*
> (Schrotter & Hürzeler, 2020, p. 102)

CHAPTER 2
TOWARDS URBAN DIGITAL TWINS

For the City of Zurich Digital Twin, another element of the build involved baseline reality capture, which was completed using a Wingtra Drone and Oblique Sony a6100 payload flying 800 ha (1980 ac) of the city centre at a GSD of 3.1 cm (1.2 in) in 6 flight hours. In Figures 2.9 and 2.10, we can see geo-information as a mesh and as a semantic model of CityGML. As Juho-Pekka Virtanen et al. state, CityGML is a critical format for exchanging large-scale urban 3D models, and CityGML has emerged as a widespread data ontology (Virtanen et al., 2021, p. 138, Noardo et al., 2021).

Thus, one of the first problems of urban digital twin projects is the form of virtualisation of a physical place, the frequency of baseline 'reality capture,' and the variety of survey instruments required for this task. This virtualisation varies across countries, with national mapping projects, municipalities, and city contracts for regular collection, and ad hoc services for specific projects. The surveying of environments is but one aspect of digital twin construction, with choices and decisions around data aggregation, standards, data

Figure 2.9
Wingtra AG, City of Zurich Mesh Model using a Wingtra Survey Drone. 800 ha (1980 ac) of the city centre at a GSD of 3.1 cm (1.2 in) in 6 flight hours, 2021.

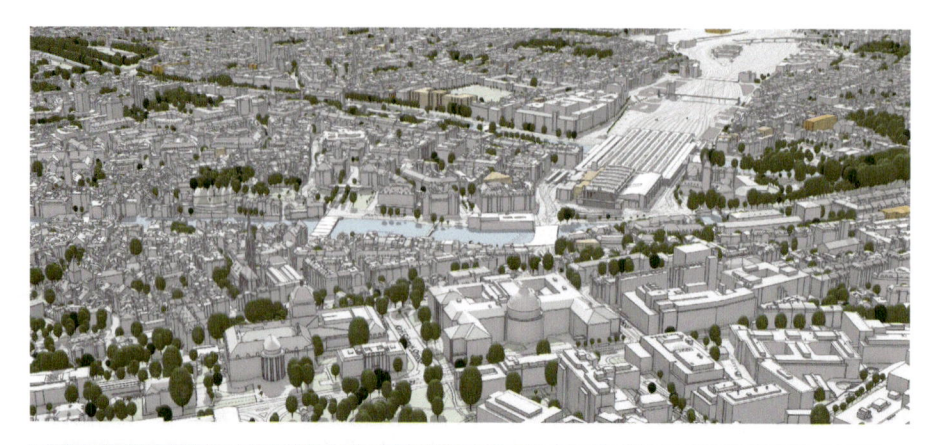

Figure 2.10
City of Zurich, Zürich Virtuell. Source: City of Zurich, Switzerland. Developed by the Municipal GIS Center of Geomatics + Vermessung in close cooperation with the Office of Urban Design and Zürich Physical Model. Scale 1:1,000, Museum of the City Archive, Office of Urban Planning. Open Governmental Portal: https://www.stadt-zuerich.ch/ted/de/index/geoz/geodaten_u_plaene/3d_stadtmodell.html.

ontologies, and the forms of storage alongside security. Cloud computing offers a range of potentials to capture large baseline areas, but such reality-capture surveys and semantic models must be realised in an accessible open data mode. The challenges of collecting baseline information highlighted here are just one issue of

CHAPTER 2
TOWARDS URBAN DIGITAL TWINS

UDT systems. Other models may run concurrently or be connected with the baseline; thus, CIMs as a precursor for UDTs are critical for validating more complex UDT systems (Omrany et al., 2022). CIMs act as a baseline for further stacking sensors and models, such as air quality, temperature, and transport, amongst many available options.

UDTs are complex and their design and purpose pose several problems, but these can potentially be overcome through geodesign frameworks to develop consensus. Geodesign, as a collaborative, deliberative, consensus-based framework, is a procedural process for decisions and lends itself to issues of UDT complex systems. Geodesign is highly participatory and particularly important here as a method for designing future UDTs due to their specificity and individuality while at the same time reflecting national ecosystems and standards.

In the case of Figure 2.11, CoExist2, for Virtual Gothenburg, the DT application revolved around testing human avatars and autonomous vehicles with a 3D environment and 3D sound for real-time traffic simulation for various time zones. Regarding its relationship with

Figure 2.11
Coexist2, Autonomous Vehicle Testing & Pedestrian Simulation, Virtual Gothenburg, 2022.

the outline Digital Twin Consortium schematic previously mentioned (Figure 2.4), a prior DT system of systems is in place to undertake divergent work (yellow and blue). Another pilot considered the transport systems of the central district, Skyfall, a flood risk and mitigation simulation, the capability of game engines to foster dialogue with planners and an AR prototype for urban design options. The project also increased citizen engagement and education by featuring the digital game Minecraft. Pilot studies are suitable for addressing DT's applications as complex systems. However, many internal challenges would also be present in creating the system and developing the best mechanisms to support research and delivery. It is essential to state that each urban digital twin should form its framework to reflect urban governance priorities and identify value-added insights possible from digital twin systems. In a geospatial context and the relationship of geo-information with UDT, systems could be conceived as multi-twins: a cluster of separate geospatial 'connected' datasets covering the extent of the applied context. This multi-twin element is best evidenced by the Federal Office of Topography Swisstopo, which provides a range of national services, all informing a clustered baseline of a DT.

For the DUET project (Figure 2.12), a collaborative DT EU-funded project testing in Athens, Greece, Pilsen, Czech Republic, and Flanders, Belgium, a data broker framework is suggested that conceptualises digital twin systems as the synthesis of existing mature technologies for applications in transportation planning and wellbeing. DUET aims to understand and develop DT ontologies and integration for several planning policy objectives. For the Flanders Regional Mobility Plan and the Flanders Environment Plan, Pilsen supports regional and local mobility and environmental policies in Flanders and, in Athens, co-creates digital services and generates decision-making approaches using common standards for greater interoperability of digital tools for policy development efficiency. In addition, DUET is exploring the system of system approaches for applications for evaluating social behaviours (Raes et al., 2021). These examples offer frameworks and schematics for design and implementation across cities, highlighting challenges and diversity

CHAPTER 2
TOWARDS URBAN DIGITAL TWINS

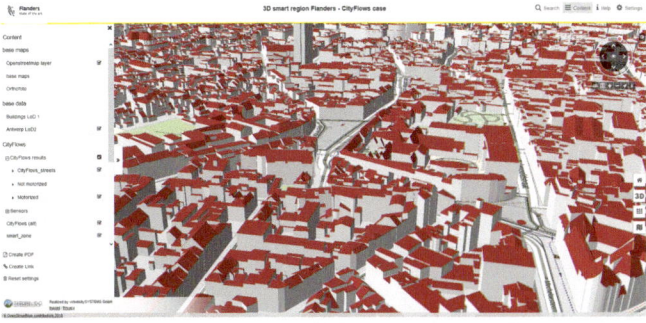

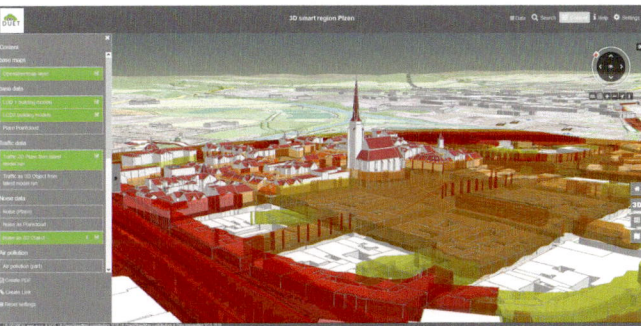

Figure 2.12
Digital Urban European Twins – DUET. Co-ordinator, Lieven Raes. Images comprised of the Flanders and the City of Ghent, an integrated regional/ local traffic model, Pilsen Digital Twin will focus on the interrelation between transport and noise pollution in a 3D environment, Athens, Digital Services An adjacent project for Antwerp city visualized flows of cars, pedestrians and cyclists. EU Horizon 2020 research and innovation programme under grant agreement No 870697. https://duet.virtualcitymap.de/.

111

of approaches. Geodesign processes for stakeholder engagement (Chapter 3) are particularly relevant for designing future complex DT systems. In each case, the success of projects has been achieved by each having a unique identity for UDT implementation and ecosystem.

Standards of Systems

OGC – 3D-IoT Platform for Smart Cities Engineering Report http://docs.opengeospatial.org/per/19-073r1.html

OGC – OGC SCIRA Pilot Engineering Report https://docs.opengeospatial.org/per/20-011.html

Industry Foundation Classes (IFC) for data sharing in the construction and facility management industries – Part 1: Data schema https://www.iso.org/standard/70303.html

ISO 12006-2:2015

OGC Urban Digital Twins Interoperability Pilot https://www.ogc.org/initiatives/ogc-urban-digital-twin-interoperability/

Building information modelling and other digital processes used in construction – Methodology to describe, author, and maintain properties in interconnected data dictionaries https://www.iso.org/standard/75401.html

Building construction – Organization of information about construction works – Part 2: Framework for classification https://www.iso.org/standard/61753.html

ISO 23386:2020

URBAN ANALYTICS

UDTs are intended as a single source of truth, but this term is highly misleading. In a geospatial context, this is difficult to define as this can involve consolidated data, and the range of quality checking and interoperability can be complex. Combining datasets without knowledge of provenance can be challenging, such as combining digital surface models with building models. Both data sets may be correct and within a set margin of error, but this may result in 'floating' elements that are strange in a visualisation, for example. This consolidation process can raise questions about the reliability of information used for various analyses and is related to issues of trustworthy AI, machine learning, and the use of large language models in urban planning. The pace of change is particularly difficult to map, though academic researchers and organisations may wish to use ALTAI – The Assessment List on Trustworthy Artificial Intelligence – as a basis to evaluate the AI elements of their DT project. The single source of truth term is problematic for UDTs regarding geospatial analysis. Urban analytics is defined as,

> *the practice of using new forms of data in combination with computational approaches to gain insight into urban processes. Increasing data availability allows us to ask new and often complex questions about cities, their economy, how they relate to the local and global environment, and much more.*
> (Singleton et al., 2018, pp. 15–16)

Urban informatics studies how people utilise and apply data in the built environment. Xintao Liu, Wenzhong Shi, and Anshu Zhang provide clear explanations of urban informatics's remit and terms.

> *Urban informatics synergizes urban science, geomatics, and informatics. Urban science reveals the principles in the urban physical environment, people's activities and flows, and their interactions. Geomatics enables better acquisition and management of geographically rich urban data. Informatics develops the technologies for utilizing the revealed principles and geographically rich urban data to develop applications and solutions to urban issues.*
> (2021, pp. 395–396)

For example, UDT analytics can be misunderstood, and there are differences between the DT system's ability to conduct simulation and prediction. DT simulations are complex calculations based on mathematical models such as agent-based modelling or computational fluid dynamics for 'airflow' simulations of an urban block; predictions are viewed as a source of (reliable) truth for a particular point in the future, such as 'what-if' scenarios. As with all algorithms, there must be trust and a clear margin of error. As Singleton and Spielman state,

113

> *Many new algorithms used to create predictions rely on big data that are used to train models, which is the process by which an algorithm learns from the past to make new or future predictions. However, in doing so, an analyst has to be certain that there are no systematic biases in such data, and that any measures taken are likely to be stable over time.*
> (Singleton & Spielman in Liu et al., 2021, p. 238)

Considering digital twins, urban capabilities, and emerging algorithms, the range of applications and analytical possibilities has vastly increased. The term single source of truth can also be misleading for decision-makers. The predictions that result from analytical tasks in a UDT should be viewed as having a statistical probability and margin of error. Filip Biljecki et al. have conducted a taxonomy of analytical possibilities and use cases using 3D GIS, indicating the evidence bases that can be generated for DTs (Biljecki et al., 2015; Biljecki, 2020). However, as Löfgren and Webster state,

> *there is an assumption that data will flow readily between public and private sector actors, but in practice, the result is that the control of the design, analysis and subsequently usage, will be in the hands of commercial actors... in data analytics.*
> (2020, p. 9)

Commercial actors' roles are not an issue per se, but the contested area is the transparency and participatory elements required from urban data analytics, and many open data models of spatial digital twins are promoted. Predictions and simulation results are often published using geospatial web platforms, but the processing and training methodologies are often kept in-house for security or commercial sensitivity. Most existing urban digital twin projects are part of national research-funded projects and have open data at the heart of UDT design, which is beneficial, but our basis of decision-making and processing should also be part of this transparency for working with and hosting urban data. Thus, UDTs' analytics is fundamentally related to aspects and challenges of smart-governance data-informed decisions and critical to participation, working with various actors and exploring socio-technical relationships (Nochta et al., 2021). This has been demonstrated in the Smart-Cambridge, UK pilot project and the West Cambridge Campus Digital Twin Demonstrator by the Institute of Manufacturing, which used reality capture, 3D modelling of the building, and over 60 sensors for monitoring and control (Qiuchen Lu et al., 2020) (Figure 2.13). In addition, a fictional city called Sunford was created to explore utility risk and mitigation of flooding in various scenarios in the CReDo project

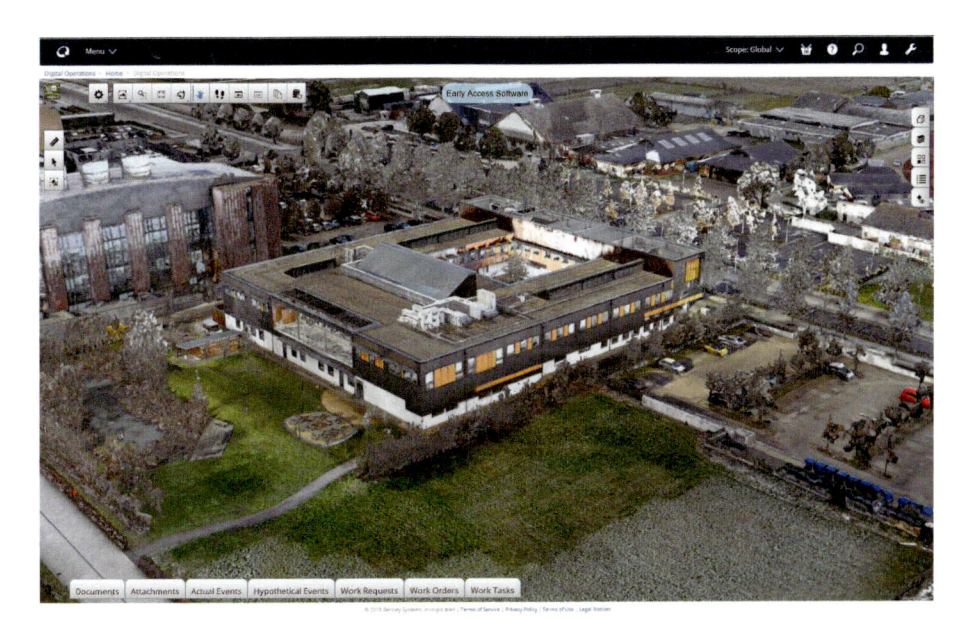

Figure 2.13
West Cambridge Development Digital Twin, Institute for Manufacturing, University of Cambridge West Cambridge campus 2017. Leads: Dr Ajith Kumar Parlikad, Reader in Asset Management, Institute for Manufacturing, University of Cambridge; Dr Richard Mortier, Professor in Computing and Human-Data Interaction at the University of Cambridge Computer Lab; Dr Ian Lewis, Director of Infrastructure Investment, University of Cambridge. Team: Dr Nicola Moretti, Dr Xiang Xie, and Dr Jorge Merino, Research Associates, Institute for Manufacturing, University of Cambridge; Dr Matthew Danish, Research Associate, Department of Computer Science and Technology, University of Cambridge; Justas Brazauskas, Research Assistant; Rob Bricheno, Senior Network Systems Specialist, University Information Services, University of Cambridge; Dr Rohit Verma, Research Associate, and Vadim Safronov, a postgraduate researcher, both based in the Systems Research Group, Department of Computer Science and Technology, University of Cambridge.

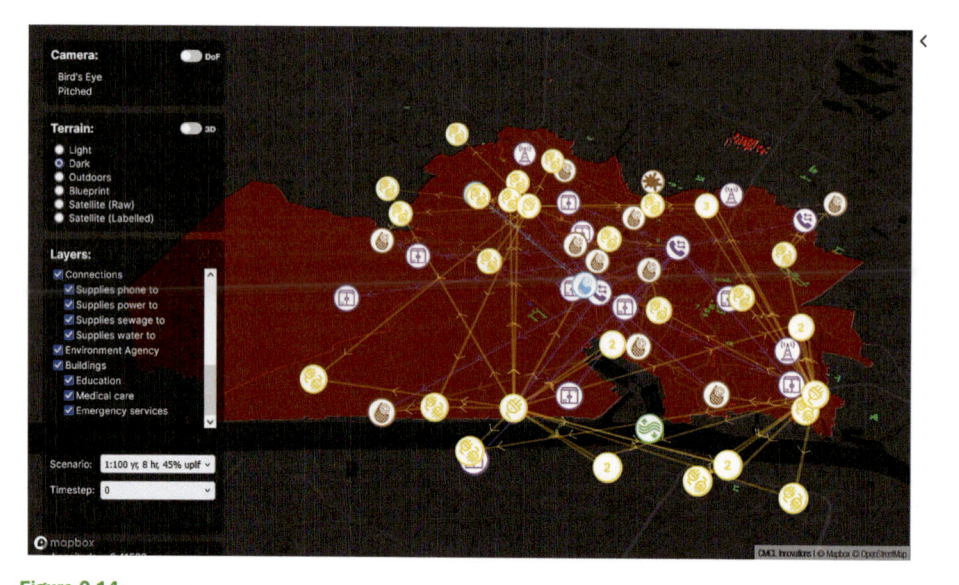

Figure 2.14
CReDo, The National Digital Twin Climate Resilience Demonstrator, 2022. CReDo displays synthetic data and infrastructure assets and connections. Various scenarios can be formed due to flood risk to measure the impact and operations of assets. CReDo was delivered through a collaboration of research centres (Universities of Cambridge, Edinburgh, Newcastle, and Warwick along with the Science and Technology Facilities Council and the Joint Centre of Excellence in Environmental Intelligence) and industry, funded by BEIS, the Connected Places Catapult, and the University of Cambridge. https://digitaltwinhub.co.uk/climate-resilience-demonstrator-credo/.

(Figure 2.14) (Hayes et al., 2022). Using a web scene, the resilience planner allows users to visualise a particular infrastructure decision's impact on the city.

In another case, Uppsala created a digital twin, Sweden, to devise a suitable development plan for city expansion for 2050 as part of a climate roadmap for NetZero production (Figure 2.15). The model utilises analytical tools in Arc GIS Urban and CityEngine to create scenarios to evaluate whether housing targets could be met using parametric modelling, and various housing targets and procedural layouts could be generated. Uppsala had already commissioned several other virtualisations of the city for personal rapid transport (PRT) from 2009 to 2012, exploring virtualised cities with scripted pedestrian bots and relationships with stakeholders in early pioneering work (Videira &Lindstrom, 2012). Scenarios feature from CIMs

Figure 2.15
City of Uppsala, 2021, Sweden Land Use, Zoning Expansion Plan, 2050.

and The Town of Morrisville, a Raleigh suburb in North Carolina. The urban planning and geospatial firm Houseal Lavigne utilised CityEngine to create two scenarios based on public consultancy that helped form the land use plan (Figure 2.16). Houseal Lavigne created a web scene for consultation but also used Unreal Engine to simulate the scenarios.

An increasing availability of pre-trained deep learning models for image extraction, pre-trained models for labelling, and classification of various geospatial elements is coming to the fore (Koumetio Tekouabou et al., 2021). The range of models applies to both still and video media such as object tracking, automated point cloud classification, and image redaction (for street-level imagery, for example). The range of tools and models becoming embedded with geospatial software has democratised skillsets in what was previously extremely specialist work and reducing manual and analytical tasks. For example, a Harvard Graduate School of Design project led by Charles Waldheim (2022) examined Ed Ruscha's photographic archive from the Getty Institute, deploying generative adversarial neural (GAN) networks to create various imaginaries based on the collection for the city of Los Angeles influenced by the AI data sculptures of Refik

Figure 2.16
Town of Morrisville and Houseal Lavigne, 2020, Morrisville Land Use, North Carolina, USA. A 2007 town centre vision was established in 2007, and the town engaged Houseal Lavigne to create a digital twin of the area to help develop community consensus. The results were an immersive experience of two development scenarios.

Figure 2.17
Chunfeng Lu, 2021, A Glimpse of a Pleasure Garden, MLA22, Charles Waldheim Studio, "Shading Sunset: Reimagining the Streets of Los Angeles for a Warmer Future", Harvard University, Graduate School of Design.

Anadol (Figure 2.17). The range of analytical approaches and GIS techniques is highly dependent on the design of the UDTs as a complex system; the development of transparent, open data; and trustworthiness in the use of AI tools that generate results, scenarios, simulations, and predictions. Urban futures in light of climate change and extreme climate events will increasingly rely on these capabilities, and a wide variety of deep learning modes can yield critical insights into future environments (Pollastri et al., 2017).

DASHBOARDS

UDT analytics rely upon clear and efficient communication of results for human-centred responses. Data dashboards have become vital for this communication and are an element in government transparency and open data ambitions of various cities and regions. For example, ESRI and Ensheng Dong from John Hopkins University created a COVID-19 dashboard which visualised the epidemic. An urban digital twin does not have to contain a dashboard; however, many smart city projects have used the tool as a connected data chain's end and public-focused element. Fundamentally, connected to dashboards is the idea of open data, which is intertwined with smart city concepts, and thus digital twins of cities and regions, though several barriers can be faced in implementing open data and engaging stakeholders (Ma & Lam, 2019). Fátima Trindade Neves et al. (2020) proposed an Open Data Impact for Smart Cities (ODISC) framework through a randomised control trial to evaluate and monitor open data initiatives in smart cities. The experiment contains a process proposed by Haynes et al. (2012). The Open Data Institute has also produced several guides and frameworks for DTs. Often standalone or hosted as part of geospatial web cloud services, dashboards are essential in meeting the 'real-time' ambition of connected spatial digital twins. As Samuel Stehle and Rob Kitchin state,

> ***Real-time data and their usefulness in a city dashboard context are then highly mediated by the technologies which collect and digitize them, expose them to analysis, and display them graphically.***
> (Stehle & Kitchin, 2020, p. 345)

In the study conducted, Stehle and Kitchin noted how mediated elements raise visual communication issues, noting the aspect of 'change blindness,' that is, the failure to attend to changes in animated transitions and the degree to which a dashboard element refreshes raises questions around principles of dashboard design (Stehle & Kitchin, 2020, p. 360). Indeed, dashboards as informatics have the potential to mislead viewers through inappropriate chart styles, refresh rates, attributed sources, and graphical features, which may result in biased or unclear results. Dashboards are visual data elements that attract the same problems as cartographic practices and the presentation of visual information, as raised by authors such as Edward Tufte (1983). Cartographic techniques and data visualisation are "the representation and presentation of data to facilitate understanding" (Kirk, 2016, p. 19), and dashboards should undergo sample testing with stakeholders. In the design field, 'personas' are often developed to represent specific groups and test the viability of a digital service or product alongside A/B tests. The visualisation of GIS information has moved from static representations of vector and raster maps based on geodatabases held on local networks or single computer units to cloud-based repositories that can align other spatial datasets with ease, allow dynamic updates from physical sensors, and increase access and dissemination of information on a multitude of platforms. Peter Hemmersam et al.'s (2015) study of participation in planning utilised DynaPlan for the Bjørvika urban development, Norway, with Masterplan by the firm SLA. In the visualisation and dashboard model, the intervention demands are based on the type of actors engaged in the development, allowing the identification of particular interests or advocacy groups involved in the planning process over a period of time. Not only does this deliver transparency to planning processes, but it fundamentally weighs who is involved in projects and for what reasons.

Dashboards incorporate various sub-sets methods of data presentation such as pie charts, doughnut charts, bar, scatter, or box plots, dot charts, stacked bars, tables, graphs, word clouds, histograms, and radar plots, amongst many others (Frost et al., 2021, Part 2). Examples include popular programs such as R and Tableau, and in-built GIS software layout features allow various visualisation options (Figure 2.18). Boston has its city-data dashboard for urban 'health,' which shows how government services are meeting targets and daily performance on a range of indicators.

Another example is the Dublin Dashboard (Figure 2.19), developed by Maynooth University following user interviews and differential

CHAPTER 2
TOWARDS URBAN DIGITAL TWINS

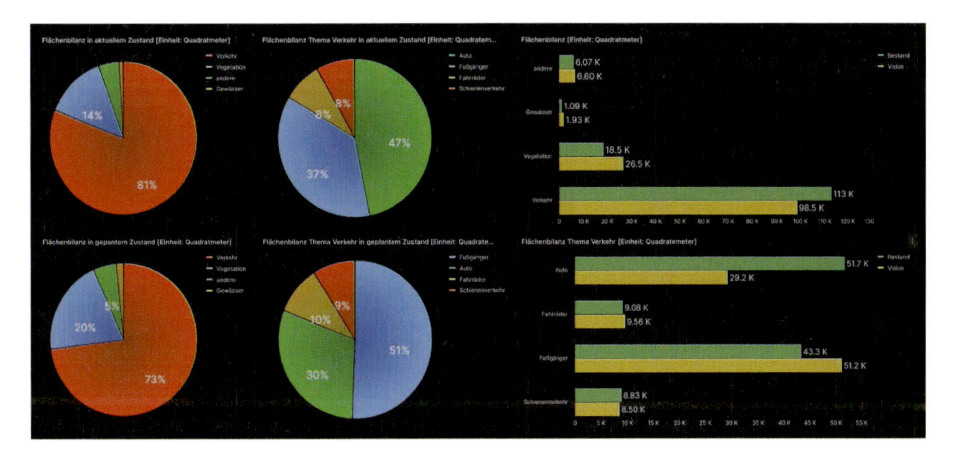

Figure 2.18
City of Munich, Digital Twin Munich, 2023, Dashboard of indicators for bicycle lane widths and area balance. https://muenchen.digital/projekte/digitaler-zwilling/01_public-participation-en.html.

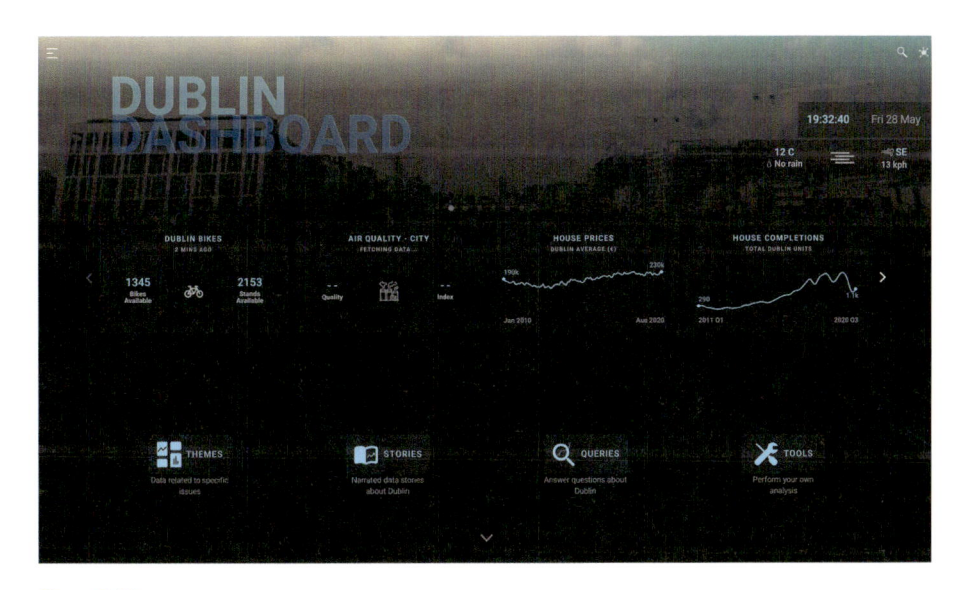

Figure 2.19
The Building City Dashboards project, PI: Rob Kitchin, funded by Science Foundation Ireland, 2016-to 2020, Maynooth University. Project Partners, Fingal County Council, Cork City Council, Cork County Council, Dublinked (which includes the two other Dublin local authorities), the Central Statistics Office (CSO), and Ordnance Survey Ireland (OSI). https://dashboards.maynoothuniversity.ie/exhibition/.

data visualisation to map optimum solutions. As Heike Vornhagen et al. state, data dashboards are critical elements that need to be carefully tuned to citizens, including the level of 'trustworthiness' of information,

> *Once a city authority has decided on the overall purpose of the dashboard, other areas such as how to engage citizens, being accountable and how to communicate this purpose through design, need to be carefully considered as these areas will contribute to the sense-making process.*
> (2021, p. 224)

Government and geospatial web platforms and dashboards are part of broad digitalisation drives. This digitalisation can be seen in the Sydney dashboard, which matches a 20-year planning strategy and its delivery to these goals (Barns, 2018, p. 9). These services are essential bridges for highly technical connected digital twin systems and social responses and the reception of analytical and performative measures. The role of dashboards will likely increase as various urban environments develop further connected systems and UDT capabilities.

Urban Dashboards	
Open Data Institute	https://theodi.org/insights/impact-stories/digital-twins-virtual-versions-of-real-world-assets/
Sydney Dashboard	https://citydashboard.be.unsw.edu.au/
Boston City Score	https://www.boston.gov/innovation-and-technology/cityscore
City of Dublin Dashboard	https://www.dublindashboard.ie/

DIGITAL TWIN INTERACTIONS

Urban digital twin interactions need to be carefully designed and implemented. The consideration of participatory modes, including digital twins for participation, will be covered in more detail in Chapter 5, which discusses gamification and the application of game techniques to non-game contexts. Serious games have emerged for urban planning, that is, games technology and methods applied to urban systems for non-entrainment purposes (Poplin, 2011). If we consider the totality of the sections in this chapter, virtualised baseline, design of complex systems, analytics, and dashboards, later phases of a connected UDT need to consider outcomes and applications, that is, the applied social context of the UDT. The nature of applied work needs to consider the participatory element of a UDT. Several options exist from the current workflows, technologies, and currently available processes. Primarily, augmented reality and serious games can offer simulative experiences for participation, as Beattie et al. in a speculative games project Maslow's Place state,

> **Visioning, foresight, or speculative urban design exercises can reveal the values and tacit and latent needs of stakeholders through discussion and experimentation that are conducive to building mutual understanding, networks, and relationships between participants.**
> (Beattie et al., 2020, p. 143)

Andrew Hudson-Smith and Valerio Signorelli, CASA, University College London, designed an interactive platform called ViLO (Figure 2.20) with the first iteration of Queen Elizabeth Olympic Park, a sporting complex built for the 2012 Summer Olympics and the 2012 Summer Paralympics in which the building's semantic features are maintained. It was built with OpenStreetMap and Mapbox API and used the CITYGML standard. Research investigations focused on mobility datasets based on Transport for London's API. Thus, bus networks, bike-sharing, and tube train services could be visualised in near real-time alongside a weather API that changed the virtual environment. Revisiting Virtual Helsinki In the Kalasatama area, the research team explored the planning and construction of digital interoperability, land and blue infrastructure

Figure 2.20
Andrew Hudson-Smith, Valerio Signorelli, ViLO Virtual London, CASA, 2017. University College London – Centre for Advanced Spatial Analysis.

options, and scenarios. The project also developed engagement through a browser OpenCities application (now Bentley systems) for augmented reality use on smart devices, which visualised development and used a public participation GIS survey for responses (KIRA-Digi Pilot Project, 2019). Again, revisiting Virtual Gothenburg, GIS assets were gamified for augmented experiences as a prototype by a student in Malmö University to research the participatory possibilities of the method aiming to increase trust and generate local values and identities of places (Klefbom, 2021, p. 44) (Chapter 1). The AR of Gothenburg Digital Twin allowed users to view, interact, and experience overlays, commentary, and histories for shared experiences of a small district, Ringön. The Gothenburg case demonstrates the work's potential and scaling to cover large geographic areas.

The participatory nature of AR and VR simulations of UDTs is an emerging field with some potential, as demonstrated by the game's agency Zoan, which developed a metaverse of its central area for virtual tourism alongside a virtual concert for the 2020 May Day celebrations, which attracted 1.4 million virtual tourists. Tangential developments of city-building games, beginning with Micropolis and SimCity and moving to contemporary games such as Cities Skylines, highlight issues faced with urban planning by non experts but also develop understanding and interactions generating pedagogic benefits (Chapter 5). These benefits include reflecting the inherent bias of the player in how a city should be formed, which can translate to serious game platforms derived from connected real-world urban environments. Urban planners and the public benefit from recreational fictional city-building games by understanding urban challenges, interactions, and interventions required by embedding themselves in the problem space. These skill sets transfer to real-world applications. However, Bradley Bereitschaft states these games are social simplifications and accept unsustainable planning practices as part of game progressions, such as ever-increasing highway infrastructure and placeless auto-centric development (2016, p. 58).

In comparison, UDTs in applied contexts for planning and participation utilise game engines to generate AR/VR experiences of future simulations of particular scenarios such as flooding, transportation, or large masterplans. Games technology and methodologies are increasingly crossovers in the domains of contemporary GIS in terms of workflows. However, GIS professionals require training in game design elements, such as user experience (UX), the elements of games – storying, worlds, components, and mechanics of games. For example, Kevin Werbach and Dan Hunter (2015) describe in their gamification toolkit the need to consider dynamics, which are the elements that exist at a high level and provide motivation through features like narrative or social interaction. Mechanics are the elements that drive player involvement and include aspects like chance, collaboration, and or rewards. Finally, components are specific examples of higher-level features: elements such as points, virtual goods, or explorations. Thus, GIS is gamified, and due consideration is required for this element of a UDT complex system.

UDTs are an emerging field, and the thematic areas of virtualisation, designing complex systems, analytics, dashboards, and interactions are intended to show the broader context and connections of applied GIS in urban planning – the companion tutorials for this chapter highlight some software workflows in establishing these elements but are not exhaustive. The next chapter will discuss geodesign tools for engagement and participation. Chapter 3 examines a range of tools and approaches for consensus building, co-creation, and community participation in urban planning and aims to overlap this section by identifying digital twin implications. The increasing range of geospatial data available alongside various urban digital twin projects and roadmaps for establishment means that the first challenge of virtualising places will be increasingly possible as various large-scale survey projects and open data ambition are realised. This virtualisation will be an essential milestone for urban futures, and digital twins can be democratised in other regions over typical ICT-centric global cities. Critical to the role of UDTs is demonstrating transparency in how they are used for decision-making and public participation alongside trustworthiness in the analytic systems.

CASE STUDY: VIRTUAL HELSINKI

Selecting the appropriate range of public-facing results from analytical activities is critical for DT success. For Virtual Helsinki, the analytical phases used Lady Bug Tools in Grasshopper, a set of open data tools for environmental design (https://www.ladybug.tools/) to conduct a wind analysis and create a series of wind roses, which is connected to weather data (https://energyplus.net/weather). Following the analysis, the acquired semantic model was imported to CityEngine and then simulated through Ansys Discovery; alternatives include (Airflow Analysis), which identified higher velocity channels in between tower blocks causing wind channels and changes to the urban microclimate. In addition to this, a solar analysis was undertaken for Kalasatama Park (Figure 2.21).

Later phases of the project for Virtual Helsinki resulted in an Energy and Climate Atlas – https://kartta.hel.fi/3d/atlas/#/ which hosts a range of tools for city council NetZero targets such as the solar potential of city buildings, heating demand of buildings, the geo-energy potential of the urban landscape, and energy performance data of buildings all held on a cloud platform. One detailed analysis focused on cold air follows for planned buildings using "meso- and micro-scale climate models, the obstacle effect of planned buildings could be determined and compared with the current development" (Schrotter & Hürzeler, 2020, p. 108). This tool allows the public to explore their property, neighbourhoods, and significant planned schemes to consider energy efficiencies and resilience of structures for NetZero goals.

All data sets can be found here: https://hri.fi/data/en_GB/dataset/helsingin-3d-kaupunkimalli

https://www.hel.fi/en/decision-making/information-on-helsinki/maps-and-geospatial-data/helsinki-3d

GEODESIGN, URBAN DIGITAL TWINS AND FUTURES

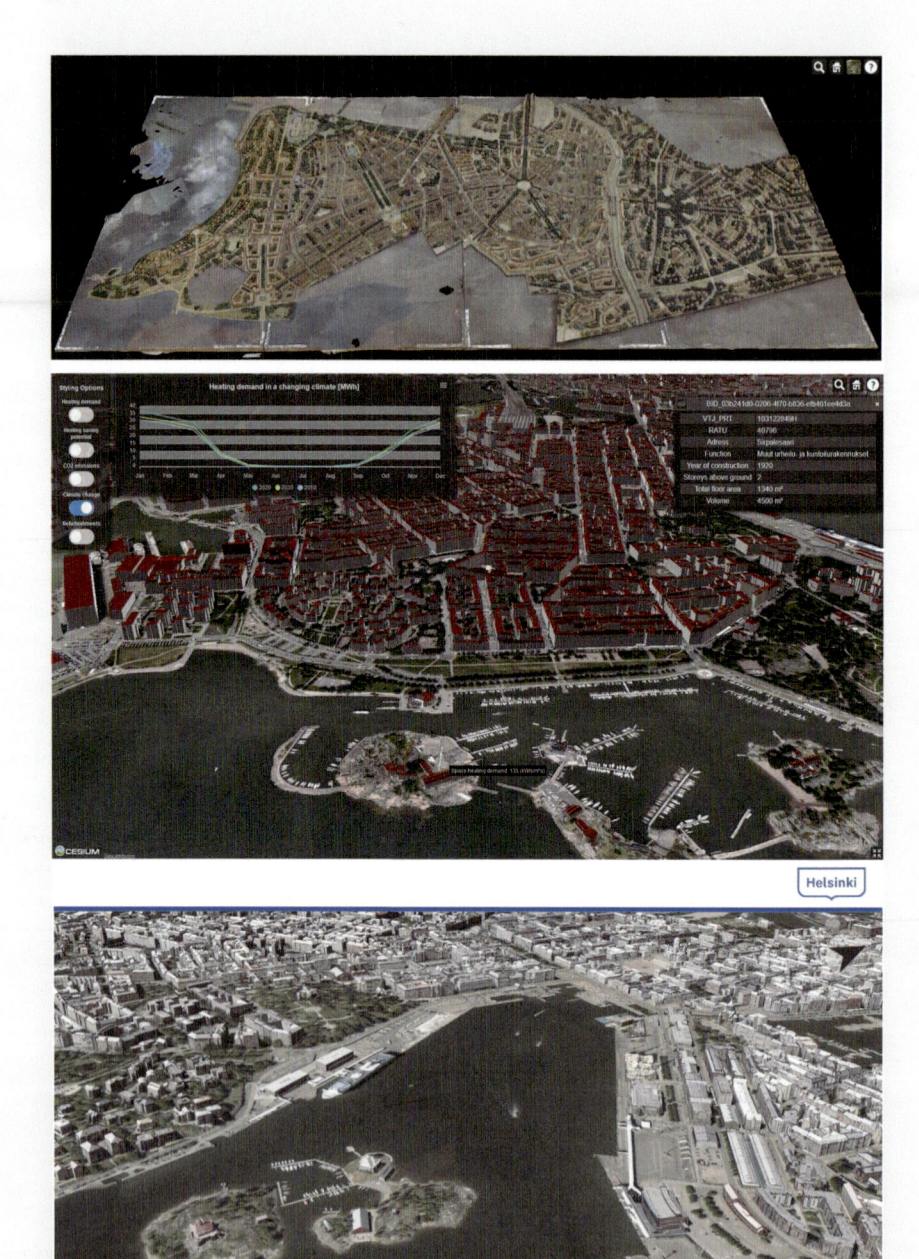

Figure 2.21
Munkkiniemi-Haaga plan, Virtual Helsinki. A miniature model of Eliel Saarinen's Munkkiniemi-Haaga plan from 1915. Solar Energy Potential, Virtual Helsinki, Hakaniemi district, 2019. Energy and Climate Atlas.

REFERENCES

Angelidou, M. (2017). Smart city planning and development shortcomings. *TeMA Journal of Land Use, Mobility and Environment*, 10. (Accessed 12/04/2018) http://www.serena.unina.it/index.php/tema/article/view/4032.

Arroyo Ohori, K., Diakité, A., Krijnen, T., Ledoux, H., & Stoter, J. (2018). Processing BIM and GIS models in practice: Experiences and recommendations from a GeoBIM project in the Netherlands. *ISPRS International Journal of Geo-Information*, 7(8), 311.

Barns, S. (2018). Smart cities and urban data platforms: Designing interfaces for smart governance. *City, Culture and Society*, 12, 5–12, ISSN 1877-9166. https://doi.org/10.1016/j.ccs.2017.09.006.

Batty, B. (2018). *Inventing future cities*. Cambridge, MA: The MIT Press.

Beattie, H., Brown, D., & Kindon, S. (2020) Solidarity through difference: Speculative participatory serious urban gaming (SPS-UG). *International Journal of Architectural Computing*, 18(2), 141–154. https://doi.org/10.1177/1478077120924337.

Bettencourt, L.M.A. (2024). Recent achievements and conceptual challenges for urban digital twins. *Nature Computational Science*, 4, 150–153. https://doi.org/10.1038/s43588-024-00604-9.

Biljecki, F. (2020). Exploration of open data in Southeast Asia to generate 3D building models. ISPRS annals of photogrammetry, remote sensing and spatial information sciences VI-4/W1–2020, 37–44.

Biljecki, F., Stoter, J., Ledoux, H., Zlatanova, S., & Çöltekin, A. (2015). Applications of 3D city models: State of the art review. *ISPRS International Journal of Geo-Information*, 4(4), 2842–2889. https://doi.org/10.3390/ijgi4042842.

Blair, G.S. (2021). Digital twins of the natural environment. *Patterns*, 2(10), 100359, ISSN 2666-3899. https://doi.org/10.1016/j.patter.2021.100359.

Bereitschaft, B. (2016). Gods of the city? Reflecting on city building games as an early introduction to urban systems. *Journal of Geography*, 115(2), 51–60. https://doi.org/10.1080/00221341.2015.1070366.

Broo, D.G., & Jennifer Schooling, J. (2021). Digital twins in infrastructure: Definitions, current practices, challenges and strategies. *International Journal of Construction Management*, 23(7), 1–10.

Brunsdon, C., & Singleton, A. (2015). *Geocomputation: A practical primer.* SAGE Publications.

CDBB. (2021). A survey of top-level ontologies to inform the ontological choices for a foundation data model. https://www.cdbb.cam.ac.

uk/what-we-do/national-digital-twin-programme/resources-top-level-ontologies-and-industry-data-models.

Charitonidou, M. (2022). Urban scale digital twins in data-driven society: Challenging digital universalism in urban planning decision-making. *International Journal of Architectural Computing*. https://doi.org/10.1177/14780771211070005.

Clark, J.H. (1976). Hierarchical geometric models for visible surface algorithms. *Communications of the ACM*, 19(10), 547–554.

Cureton, P., & Hartley, E. (2023). City Information Models (CIMs) as precursors for Urban Digital Twins (UDTs): A case study of Lancaster. *Frontiers in Built Environment*, 9, 1048510. https://doi.org/10.3389/fbuil.2023.1048510.

de Laat, R., & van Berlo, L. (2011). Integration of BIM and GIS: The development of the CityGML GeoBIM extension. In *Advances in 3D Geoinformation sciences*. Springer.

Digital Twin Consortium. (2023). Platform stack architectural framework: An introductory guide, A digital twin consortium white paper, 11 July 2023. https://www.digitaltwinconsortium.org/platform-stack-architectural-fram-formework-an-introductory-guide-form/.

Frost, A., Sturt, T., Kynvin. J., & Fernandez Gallardo, S. (2021). *Communicating with data visualisation: A practical guide*. Sage Publications Ltd.

García-Sánchez, C., Vitalis, S., Paden, I., & Stoter, J. (2021). The impact of level of detail in 3D city models for CFD-based wind flow simulations, 3D Geoinfo, 11–14 October 2021. ISPRS Archives.

Grieves, M., & Vickers, J. (2017). Digital twin: Mitigating unpredictable, undesirable emergent behaviour in complex systems. In F.-J. Kahlen, S. Flumerfelt, & A. Alves (Eds.), *Transdisciplinary perspectives on complex systems: New findings and approaches* (pp. 85–113). Springer International Publishing. https://doi.org/10.1007/978-3-319-38756-7_4.

Hämäläinen, M. (2021). Urban development with dynamic digital twins in Helsinki city. *IET Smart Cities*, 3(4), 201–210. https://doi.org/10.1049/smc2.12015.

Haraguchi, M., Funahashi, T., & Biljecki, F. (2024). Assessing governance implications of city digital twin technology: A maturity model approach. *Technological Forecasting and Social Change*, 204, 123409, ISSN 0040-1625. https://doi.org/10.1016/j.techfore.2024.123409.

Haynes, L., Goldacre, B., & Torgerson, D. (2012). Test, learn, adapt: Developing public policy with randomized controlled trials. Retrieved from https://www.gov.uk/government/uploads/system/uploads/attachment_data/file/62529/TLA-1906126.pdf.

Hayes, S. et al. (2022). CReDo: Overview report. Centre for Digital Built Britain. https://doi.org/10.17863/CAM.82908.

Hemmersam, P., Martin, N., Westvang, E., Aspen, J., & Morrison, A. (2015). Exploring urban data visualization and public participation in planning. *Journal of Urban Technology*, 22(4), 45–64. https://doi.org/10.1080/10630732.2015.1073898.

Ketzler, B., Naserentin, V., Latino, F., Zangelidis, C., Thuvander, L., & Logg, A. (2020). Digital twins for cities: A state of the art review. *Built Environment*, 46(4), 547–573(27).

KIRA–Digi Pilot Project. (2019). *The Kalasatama digital twins project digital twins project*. Ministry of the Environment.

Kirk, A. (2016). *Data visualisation: A handbook for data driven design*. Sage Publications Ltd.

Klefbom, S. (2021). Augmented urban values: Virtual Gothenburg as a place for citizen dialogue and shared lived experiences (Dissertation). Retrieved from https://urn.kb.se/resolve?urn=urn:nbn:se:mau:diva-45934.

Klippel, A., Sajjadi, P., Zhao, J., Wallgrün, J.O., Huang, J., & Bagher, M.M. (2021). Embodied digital twins for environmental applications. ISPRS annals of the photogrammetry, remote sensing and spatial information sciences, -4–2021, 193–200. https://doi.org/10.5194/isprs-annals-V-4-2021-193-2021.

Koumetio Tekouabou, S.C., Diop, E.B., Azmi, R., Jaligot, R., & Chenal, J. (2021). Reviewing the application of machine learning methods to model urban form indicators in planning decision support systems: Potential, issues and challenges. *Journal of King Saud University - Computer and Information Sciences*, ISSN 1319-1578. https://doi.org/10.1016/j.jksuci.2021.08.007.

Lehner, H., & Dorffner, L. (2020). Digital geoTwin Vienna: Towards a digital twin city as geodata hub. *PFG – Journal of Photogrammetry, Remote Sensing and Geoinformation Science*, 88, 63–75. https://doi.org/10.1007/s41064-020-00101-4.

Lehner, H., Kordasch, S.L., Glatz, C., & Agugiaro, G. (2024). Digital geoTwin: A CityGML-based data model for the virtual replica of the city of Vienna. In T.H. Kolbe, A. Donaubauer, & C. Beil (Eds.), *Recent advances in 3D geoinformation science*. 3DGeoInfo 2023. Lecture notes in geoinformation and cartography. Springer, Cham. https://doi.org/10.1007/978-3-031-43699-4_32.

Li, W., Michael, B., & Goodchild, M.F. (2020). Real-time GIS for smart cities. *International Journal of Geographical Information Science*, 34(2), 311–324. https://doi.org/10.1080/13658816.2019.1673397.

Liu, X., Shi, W., & Zhang, A. (2021). Advances in urban informatics. *Environment and Planning B: Urban Analytics and City Science*, 48(3), 395–399. https://doi.org/10.1177/2399808321998468.

Löfgren, K., & Webster, C.W.R. (2020). The value of Big Data in government: The case of 'smart cities'. *Big Data & Society*. https://doi.org/10.1177/2053951720912775.

Lutz, C. (2019). Digital inequalities in the age of artificial intelligence and big data. *Human Behavior and Emerging Technologies*, 1, 141–148. https://doi.org/10.1002/hbe2.140.

Ma, R.Q., & Lam, P.T.I. (2019). Investigating the barriers faced by stakeholders in open data development: A study on Hong Kong as a "smart city". *Cities*, 92 (2019), 36–46. https://doi.org/10.1016/j.cities.2019.03.009.

Noardo, F., Harrie, L., Arroyo Ohori, K., Biljecki, F., Ellul, C., Krijnen, T., Eriksson, H., Guler, D., Hintz, D., Jadidi, M.A., Pla, M., Sanchez, S., Soini, V.-P., Stouffs, R., Tekavec, J., & Stoter, J. (2020). Tools for BIM-GIS integration (IFC georeferencing and conversions): Results from the GeoBIM benchmark 2019. *ISPRS International Journal of Geo-Information*, 9, 502.

Noardo, F., Krijnen, T., Arroyo Ohori, K., Biljecki, F., Ellul, C., Harrie, L., Eriksson, H., Polia, L., Salheb, N., Tauscher, H., van Liempt, J., Goerne, H., Hintz, D., Kaiser, T., Leoni, C., Warchol, A., Stoter, J. (2021). Reference study of IFC software support: The GeoBIM benchmark 2019 – Part I. *Transactions in GIS*.

Noardo, F., Arroyo Ohori, K., Biljecki, F., Ellul, C., Harrie, L., Krijnen, T., Eriksson, H., van Liempt, J., Pla, M., Ruiz, A., Hintz, D., Krueger, N., Leoni, C., Leoz, L., Moraru, D., Vitalis, S., Willkomm, P., & Stoter, J.(Pre-print). Reference study of CityGML software support: The GeoBIM benchmark 2019 – Part II. *Transactions in GIS*, 25, 842–868.

Nochta, T., Wan, L., Schooling, J.M., & Parlikad, A.K. (2021). A sociotechnical perspective on urban analytics: The case of city-scale digital twins. *Journal of Urban Technology*, 28(1–2), 263–287. https://doi.org/10.1080/10630732.2020.1798177.

Omrany, H., Ghaffarianhoseini, A., Ghaffarianhoseini, A., & Clements-Croome, D.J. (2022). The uptake of City Information Modelling (CIM): A comprehensive review of current implementations, challenges and future outlook. *Smart and Sustainable Built Environment*. https://doi.org/10.1108/SASBE-06-2022-0116.

Pollastri, S., Boyko, C., & Cooper, R., Dunn, N., Clune, S., Coulton, C. (2017). Envisioning urban futures: From narratives to composites. *The Design Journal*, 20(1), S4365–1477.

Poplin, A. (2011). Games and serious games in urban planning: Study cases. In B. Murgante, O. Gervasi, A. Iglesias, D. Taniar, & B.O. Apduhan (Eds.), *Computational science and its applications - ICCSA 2011.* ICCSA 2011. Lecture notes in computer science, vol. 6783. Springer, Berlin, Heidelberg. https://doi.org/10.1007/978-3-642-21887-3_1.

Qian, Y., & Leng, J. (2021). CIM-based modeling and simulating technology roadmap for maintaining and managing Chinese rural traditional residential dwellings. *Journal of Building Engineering*, 44, 103248. https://doi.org/10.1016/j.jobe.2021.103248.

Qiuchen, L., Xiang, X., Kumar, P.A., Jennifer, S., & Eirini, K. (2021). Moving from building information models to digital twins for operation and maintenance. Proceedings of the Institution of Civil Engineers - Smart Infrastructure and Construction, 174(2), 46–56.

Qiuchen Lu, A. K. P. (2020). Developing a digital twin at building and city levels: Case study of West Cambridge Campus. *Journal of Management in Engineering.* American Society of Civil Engineers. https://doi.org/10.1061/(ASCE)ME.1943-5479.0000763.

Raes, L. et al. (2021). DUET: A framework for building interoperable and trusted digital twins of smart cities. *IEEE Internet Computing.* https://doi.org/10.1109/MIC.2021.3060962.

Stehle, S., & Kitchin, R. (2020). Real-time and archival data visualization techniques in city dashboards. *International Journal of Geographical Information Science*, 34(2), 344–366. https://doi.org/10.1080/13658816.2019.1594823.

Schrotter, G., & Hürzeler, C. (2020). The digital twin of the city of Zurich for urban planning. *PFG*, 88, 99–112. https://doi.org/10.1007/s41064-020-00092-2.

Singleton, A., Spielman, S., & Folch, D. (2018). *Urban analytics.* SAGE Publications Ltd. https://doi.org/10.4135/9781529793703.

Song, Y. et al. (2017). Trends and opportunities of BIM-GIS integration in the architecture, engineering and construction industry: A review from a Spatio-temporal statistical perspective. *ISPRS International Journal of Geo-Information*, 6(12), 397.

Souza, L., & Bueno, C. (2022). City Information Modelling as a support decision tool for planning and management of cities: A systematic literature review and bibliometric analysis. *Building and Environment*, 207, Part A, 108403, ISSN 0360-1323. https://doi.org/10.1016/j.buildenv.2021.108403.

Stankutė, S., & Asche, H. (2009). An integrative approach to geospatial data fusion. In O. Gervasi, D. Taniar, B. Murgante, A. Laganà, Y. Mun, & M.L. Gavrilova (Eds.), *Computational science and its* applications – ICCSA 2009. ICCSA 2009. Lecture notes in computer science, vol. 5592. Springer, Berlin, Heidelberg. https://doi.org/10.1007/978-3-642-02454-2_35.

Tufte, E. (1983). *The visual display of quantitative information.* Graphics Press.

Trindade Neves, F., de Castro Neto, M., & Aparicio, M. (2020). The impacts of open data initiatives on smart cities: A framework for evaluation and monitoring. *Cities*, 106, 102860, ISSN 0264-2751. https://doi.org/10.1016/j.cities.2020.102860.

van der Valk, H., Haße, H., Möller, F. et al. (2021). Archetypes of digital twins. *Business & Information Systems Engineering.* https://doi.org/10.1007/s12599-021-00727-7.

Vornhagen, H., Zarrouk, M., Davis, B., & Young, K. (2021). Do city dashboards make sense? Conceptualizing user experiences and challenges in using city dashboards. A case study. In *DG.O2021: The 22nd annual international conference on digital government research (DG.O'21)* (pp. 219–226). Association for Computing Machinery, New York, NY.

Videira Lopes, C., & Lindstrom, C. (2012). Virtual cities in urban planning: The Uppsala case study. *Journal of Theoretical and Applied Electronic Commerce Research*, 7(3), 88–100. https://doi.org/10.4067/S0718-18762012000300009.

Virtanen, J.-P., Jaalama, K., Puustinen, T., Julin, A., Hyyppä, J., & Hyyppä, H. (2021). Near real-time semantic view analysis of 3D city models in web browser. *ISPRS International Journal of Geo-Information*, 10(3), 138. https://doi.org/10.3390/ijgi10030138.

Waldheim, C. (2022) Shading sunset: Charles Waldheim on reimagining the streets of Los Angeles for a warmer future. (Accessed 20/03/23) https://www.gsd.harvard.edu/2021/04/shading-sunset-charles-waldheim-on-reimagining-the-streets-of-los-angeles-for-a-warmer-future/.

Wildfire, C. (2020). How can we spearhead city-scale digital twins? https://www.mottmac.com/views/how-can-we-spearhead-city-scale-digital-twins.

Werbach, K., & Hunter, D. (2015). *The gamification toolkit: Dynamics, mechanics, and components for the win.* Wharton School Press, 2015.

Xu, Z., Qi, M., Wu, Y., Hao, X., & Yang, Y. (2021). City information modeling: State of the art. *Applied Sciences*, 11(19), 9333. https://doi.org/10.3390/app11199333.

Xue, F., Liupengfei, W., & Lu, W. (2021). Semantic enrichment of building and city information models: A ten-year review. *Advanced Engineering Informatics*, 47, 101245. https://doi.org/10.1016/j.aei.2020.101245.

Zomer, T., Neely, A., Sacks, R., & Parlikad, A. (2021). A practice-based conceptual model on building information modelling (BIM) benefits realisation. In E. Toledo Santos, & S. Scheer (Eds.), *Proceedings of the 18th international conference on computing in civil and building engineering*. ICCCBE 2020. Lecture notes in civil engineering, vol. 98. Springer, Cham. https://doi.org/10.1007/978-3-030-51295-8_29.

CHAPTER 3

GEODESIGN AND URBAN DIGITAL TWINS

INTRODUCTION

Maps fundamentally have an agency in that they describe physical and material realities and are also the medium in which various future scenarios are described and decisions formed. These transformations range across natural and man-made interventions. Maps provide the fundamental data of city information models (CIMs) and 3D geographic information systems (GIS), reflecting broader issues and challenges in urban digital twins (UDTs) (Batty, 2018). Like UDTs, the map is a system of knowledge describing the complexity of the world or the representative outcome of the 'best' and most appropriate forms of modelling of an environment (Blair, 2021). From this basis, collaborative and deliberative decisions are undertaken to which tangible changes are enacted. This raises key questions about the most appropriate forms of governance of UDTs, frameworks, value chains, and structured information of complex urban environments (Caldarelli et al., 2023). UDTs can never be a standalone computational and technological exercise; fundamentally, they are socio-technical systems (Nochta et al., 2021).

The use of GIS in environmental science (Goodchild, 1992) is a key mode of scientific communication and stakeholder engagement. Roger Tomlinson is one of the originators of modern GIS. In Tomlinson's notation of the 1981 Harvard Computer Graphics Week Program, he states that it is "impossible to map the world – we select – and make graphics of it so that we can understand it" (Figure 3.1). Tomlinson's quote is representative of the early computational and graphic experiments of various computer labs, which established many principal applications of GIS and digital maps (Hessler, 2015; Longley et al., 2001).

Contemporary GIS capability and the accessibility of geospatial data provide abundant spatial measurements, and we are witnessing a period of large-scale earth models for climate change (Li et al., 2023). Geographic science is under continual technological development, for example, through 'Big Earth Data' as emphasised by

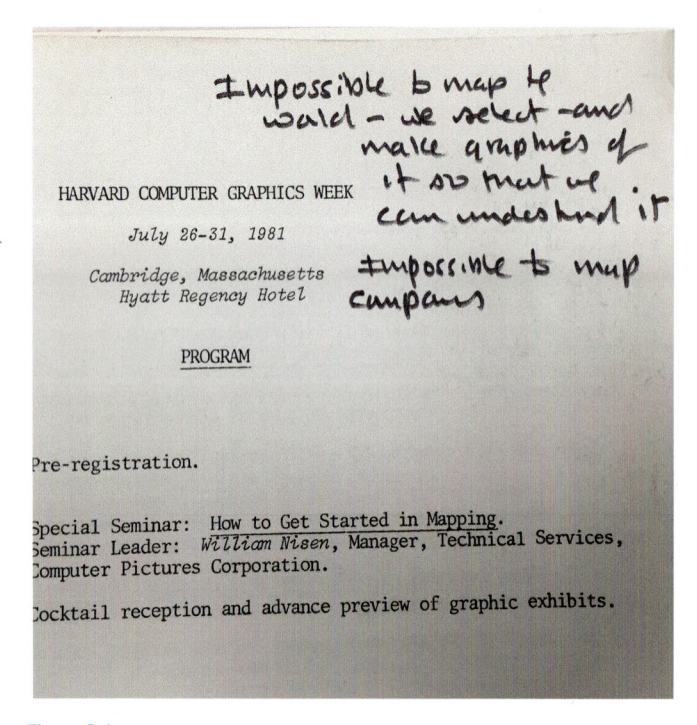

Figure 3.1
Roger Tomlinson's Notes on the 1981 Harvard Computer Graphics Week Program. "Impossible to map the worlds-we select-and make graphics of it so that we can understand it". Geography and Map Division, Library of Congress.

Earth 2 APIs by NVIDIA. At a 2-km scale, NVIDIA's AI-powered 'Weather engine' system generates simulations of the global atmosphere, cloud cover, and extreme weather (Figure 3.2). The cloud system allows various analytics and localised weather analysis for disaster reduction and future weather challenges. NIVDIA's model has simulative capability based on rigorous artificial intelligence (AI), machine learning (ML), and modelling for weather-induced geophysical hazards and climate change adaptation. This digital twin is a prime example of the socio-technical relationships and decision-making that GIS mapping and modelling packages have brought about, as decision-making feeds into flooding risk analysis and infrastructure investment, amongst other government activities (Wu & Chiang, 2018).

CHAPTER 3
GEODESIGN AND URBAN DIGITAL TWINS

Figure 3.2
NVIDIA, Earth Climate Digital Twin, 2024.

There is a broad interest in design thinking, co-design, collaborative decision-making, and multi-stakeholder engagement across environmental science, which has also been recognised as a central research paradigm. Indeed, such need for creative problem-solving can be seen from a wide variety of perspectives and tools, including The Theory of Inventive Problem Solving/Teoriya Resheniya Izobretatelskikh Zadatch (TRIZ) tools for engineering and business strategy, established by Genrikh Altshuller in 1946 (Zlotin & Zusman, 2013). From a UDT perspective, which has been discussed by Timea Nochta et al. based on the Cambridge digital twin case, who states that,

Assessing the impact of contextual characteristics on the design and implementation of digital twins for city planning (and management) is necessary to better

> *understand the risks and challenges they might pose, as well as the tailored ways to incorporate CDTs in local governance.*
>
> (Nochta et al., 2019, p. 2)

The Centre for Digital Built Britain report highlighted the need for continual socio-technical dialogue for UDTs as the systems evolve and interact over time. Decisions are often formed for the built environment in a completely different process from those for natural environments. Land use models may be used for the built environment, with CIMs or National Mapping Agency data, whereas environmental ones could be formed at the critical interface of earth systems and localised ones such as land surface models (LSMs) or ecological surveys. The difference in modelling approaches and decision-making across areas led to the Gemini Principles (2018) for information management of digital twins (DTs) and Apollo Protocol (2022) for a common language across manufacturing and the built environment DTs. UDTs are not just about data governance and data-driven systems but also the participatory aspect and local sovereignty of UDTs. These aspects are currently being explored in Hamburg's Connected Digital Twins sub-project (Figure 3.3).

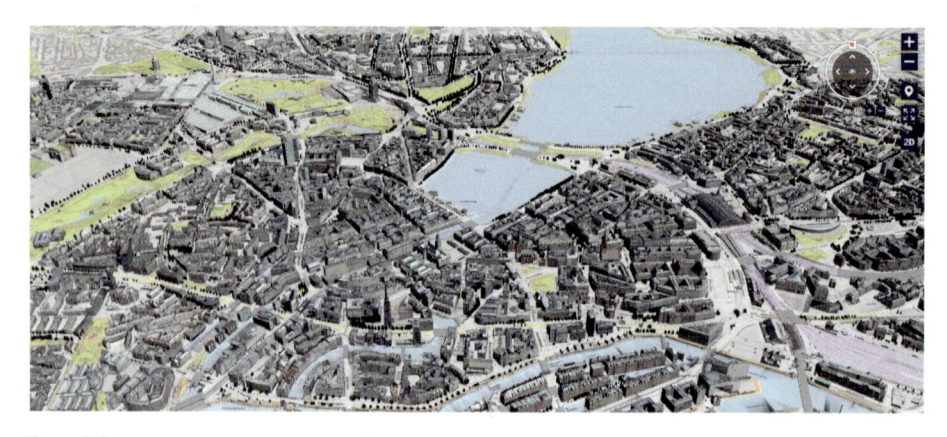

Figure 3.3
City of Hamburg 3D Model, Geo-Information as part of the Connected Urban Twins Project (Hamburg, Leipzig, Munich, Germany).

The embryonic technologies and socio-technical issues that emerge in UDTs raise questions about what frameworks and approaches are appropriate for long-term urban management.

One such socio-technical theorisation termed geodesign was created as an open, multidisciplinary approach for collaborative and deliberative decision-making for regional planning and landscape studies that emerged from the work of Carl Steinitz in 1990 (Steinitz, 1990). Geodesign, as a term, received early uptake by researchers in the process (Kunzmann, 1993) and has been in operation for over 30 years via a number of key practitioners in geodesign, including Ana Clara, Mourao Moura, Michele Campagna, and Brian Orland. Geodesign was informed by earlier paradigm-shifting theories of environments in work such as Ian McHarg, 'Design with Nature', 1969, who proposed a rationalised 'layer-cake' method for landscape assessment (Introduction). Mcharg, as we have seen, would assess suitability and land use through map overlays for comparative analysis (Herrington, 2010). Flaxman and Ervin presented the most commonly cited definition of geodesign,

> **Geodesign is a method which tightly couples the creation of proposals for change with impact simulations informed by geographic contexts and systems thinking and normally supported by digital technology.**
> (Flaxman, 2010)

The geospatial company ESRI actively promoted geodesign as a framework approached by Steinitz's former student and company founder, Jack Dangermond, naturally complementing its systems and extending beyond landscape planning and education into environmental fields (Paradis et al., 2013; Machl et al., 2019; Li & Kim, 2022; Schroth, 2023). As William Miller describes, "geodesign is design in geographic space (geo-scape). Correspondingly, the purpose of geodesign is to facilitate life in geographic space (geo-scape)" (Miller, 2012, p. 16). However, there are many 'designers' that do not necessarily utilise the Steinitz framework but do have similar interests in procedural systems and evaluation of geographic information. Since establishing the framework, nature-based solutions (NBS) research has actively utilised co-design processes and geodesign (Albert et al., 2021; Gottwald et al., 2021; Tran et al., 2023). Notably, geodesign case studies have been published by Shannon McElvaney (2012), which include visioning for Florida, USA 2050 and post-industrial transition in Los Angeles, amongst others.

The geodesign framework comprises six iterative procedural steps, which consist of assessment (3) and intervention (3) (Figure 3.4) (Cureton & He, 2023). Three phases (iterations) of the six questions below are suggested: in this interpretation of Geodesign in a playable carddeck.

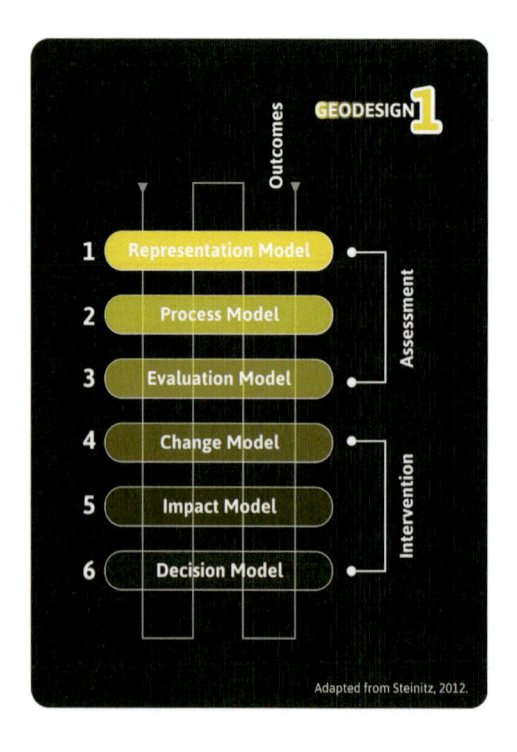

Figure 3.4
Paul Cureton, Lisha He, geodesign Phases (Cureton & Hartley, 2023). A collaborative framework is available at Figma in English and Chinese. https://www.figma.com/community/file/1253768043463157544/Geodesign-a-Collaborative-Framework.

1. **Why?**
2. **How?**
3. **What, Where, and When? (Steinitz, 2012, pp. 26–34)**

The six framework steps are intended to be repeated two or three times, asking the Why, How, What, Where, and When questions in order to enact results. Geodesign is intended to be delivered in workshops, and contemporary situations have demonstrated hybrid arrangements as well as the use of digital platforms such as the geodesign hub. Workshops can be delivered using analogue or digital modes. The process is discussed in detail in the later section, and Steinitz organised the framework based on six core research questions and what is phrased as 'models,' which is the fuzzy assimilation of data, knowledge, and assessment for each phase, which requires the specificity of the study to be set:

1. **"How should the study area be described?" (Representation models)**
2. **"How does the study area function?" (Process models)**
3. **"Is the current study area working well?" (Evaluation models)**
4. **"How might the study area be altered?" (Change models)**
5. **"What differences might the changes cause?" (Impact models)**
6. **"How should the study area be changed?" (Decision models)**

(Steinitz, 2012, p. 3)

Steinitz describes four broad stakeholder groups involved in geodesign addressing these questions, which are comprised of (1) residents and communities of the study area, (2) geographers, including economic geographers, social and natural scientists, hydrologists, and ecologists, amongst others, coupled with (3) design professionals such as architects, planners, landscape architects, civil engineers with technologists and (4) ICT support platforms (Steinitz, 2012, p. 4). The nature of the stakeholder groups originally described by

Steinitz lends itself to the multidisciplinary teams present in UDTs (Somma et al., 2022).

The various phases of the geodesign framework build upon or have similarly tangential aspects. As Foster asserts (2016), geodesign phases have divergent points to inform the process following a convergent phase to develop solutions, as seen in other design models such as Double Diamond by the British Design Council (Figure 3.5) and the divergence-convergence model of Bela H. Banathy (1996). Foster also recommends the components necessary for organising case study methods for geodesign, which has a structured sequence involving an overview, landscape performance benefits, challenge, solution, sustainability, costings, and lessons learned, which can improve deliverability (Banathy, 1996,

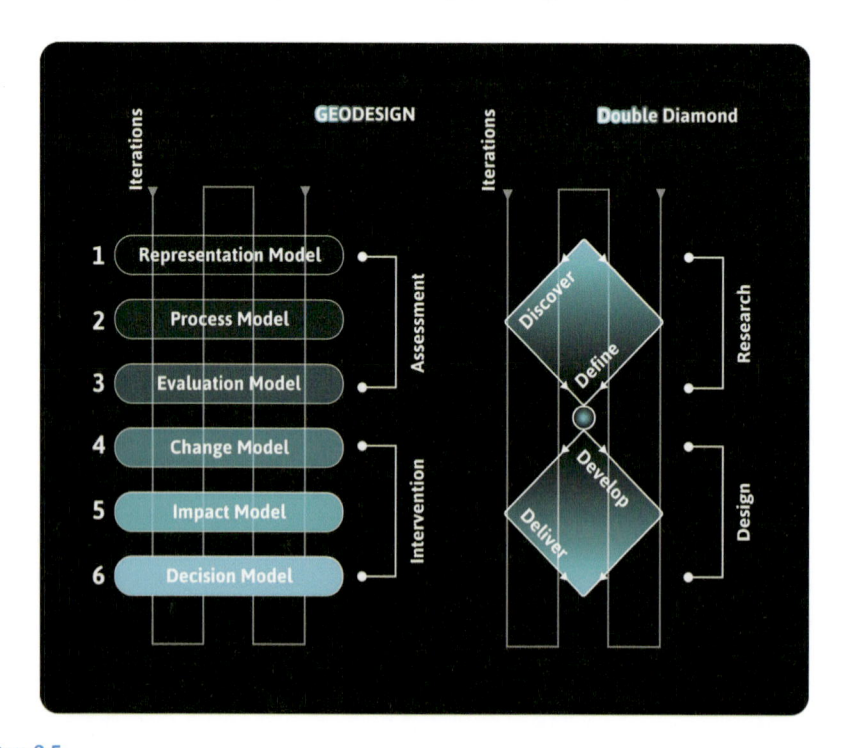

Figure 3.5
Author, Adapted Geodesign Phases by (Steinitz, 2012) and Double Diamond, British Design Council, Double Diamond, Design Process Model (2015). CC-BY.

p. 98). Kelleann Foster has detailed the context of the geodesign framework concerning other design thinking frameworks (2016). A detailed investigation of Steinitz's framework can be found in (Hollstein, 2019), which the process is examined at length and criticisms are weighted. Hollstein states that a "systematic, procedurally-arranged framework, engaging the analysis synthesis relationship" for environmental design fields is highly rationalised. However, criticism ignores the adaptability of the process and the systematic procession that the framework offers in order to solve 'wicked' problems (Rittel & Webber, 1973; Hollstein, 2019, p. 65). If geodesign processes and framework are relevant to the application of UDTs, it is worth analysing a brief historical overview of developments, the geodesign process, and a variety of geographic applications and cases relating to planning resilience.

HISTORY OF COMPUTER GRAPHICS AND GEODESIGN

Nick Chrisman's book 'Charting the Unknown: How Computer Mapping at Harvard Became GIS' (2006) discusses the Harvard Laboratory of Computer Graphics and its relationship with the design school from the early 1960s. The lab, through the direction of Howard Fisher and other directors, would experiment and release formative digital computer mapping programs such as SYMAP (SYnagrahic MAPping) program, 1963, and ODYSSEY, vector-based GIS, 1970 as well as Builder by Bruce Donald, 1981, a 3D CAD package experimenting in design and computation (Robertson, 1967; Shepard, 1984) (Figure 3.6). Carl Steinitz has accounted for this period of experimentation (2013) and the establishment of the built environment curricula (Steinitz and Rogers, 1970; Steinitz, 2013). Carl Steinitz's (2012) book, 'A Framework for Geodesign: Changing Geography by Design,' set the key geodesign concepts. Further work on the framework appeared via the ERSI Geodesign Summit and the International Geodesign

145

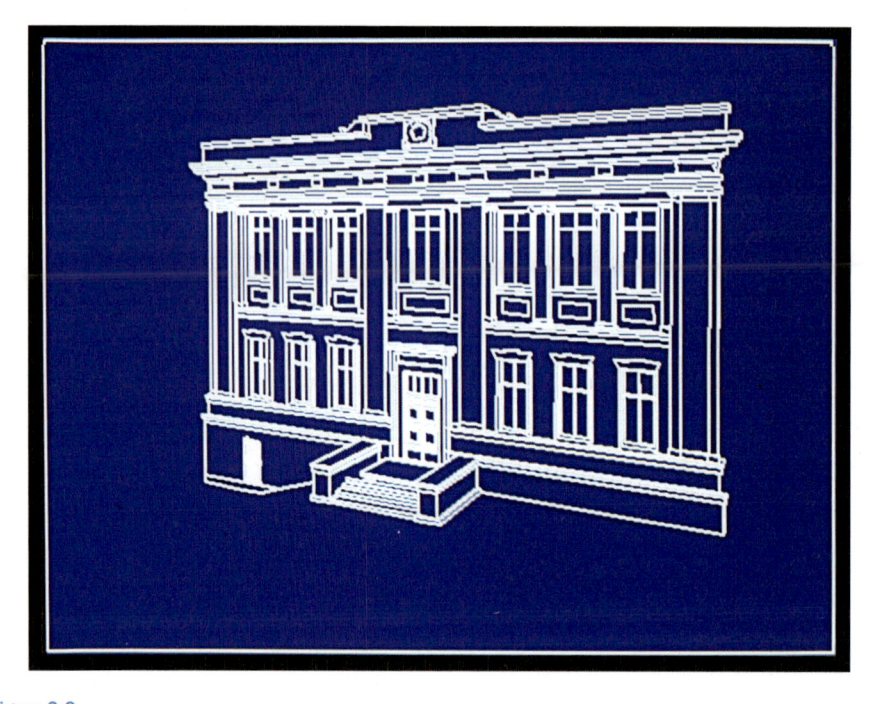

Figure 3.6
Bruce Donald with Harvard GSD M.Arch. students in a studio class, Render of Robinson Hall in Harvard Yard, former home of the Graduate School of Design, BUILDER: A Database and Display Program for Computer-aided Architectural Design. A system for representation and display of design problems in architecture and urban design. Laboratory for Computer Graphics and Spatial Analysis, Graduate School of Design, Harvard University. Cambridge, MA (1981).

Collaboration (IGC) as a worldwide academic consortium to explore scenario-based challenges (Fisher et al., 2020; Kingston et al., 2020). As geodesign emerged, Orland (2015) proposed additional phases to Steinitz's framework, including 'Storytelling, System exploration games, and Group interactions' (Orland & Murtha, 2015; Orland, 2016; Shakeri et al., 2016).

One particular aspect of geodesign, which lends itself to UDTs, is the scenario planning aspect. Steinitz's contemporary work has looked at this aspect for the Upper San Pedro River Basin in Arizona, USA, and Sonora, Mexico (Steinitz et al., 2002). As Ming-Chun Lee states, the origins of scenario planning stem from Herman Kahn at the RAND Corporation in the 1950s. Various serious games evaluated

Figure 3.7
Engineering Operations game with Milton Weiner (3rd from right), Olaf Helmer (2nd from right), and others, 1966.
Archives & Special Collections, RAND.

the effects of war and politics in which various scenarios were played out (Figure 3.7). As Mann Virdee and Megan Hughes state,

> *Scenarios are not about predicting the future. They are, rather, a rigorous and methodical way to consider several imagined future situations, or contexts, which could come to pass. Generating scenarios typically involves identifying a set of influential "drivers of change" whose outcomes may be predictable, such as demographic changes. Combinations of these drivers create a range of plausible future states.*
> (Virdee & Hughes, 2022)

Scenarios or 'what-ifs' have been fundamentally embedded in the history of GIS and regional planning (Klosterman, 1999; Lee,

2016, p. 10). Geodesign scenarios can be used to scope the study methods, outline the study, and evaluate the intervention's impacts (Steinitz, 2012, p. 40). Geodesign scenarios can potentially address Anthropocene issues in dealing with uncertainty in forecasts and practice (Minner, 2017; Shearer, 2022). As Andy-Hudson-Smith and Michael Batty state, the benefit of utilisation of the framework,

> *there may not be consensus, but the process which is highly automated using GIS technologies provides an environment for conflict resolution for continued sharpening of the set of best alternatives.*
> (Hudson-Smith & Batty, 2022, p. 1154)

Contemporary geodesign scenarios feature in many geodesign projects, such as the project by Kaihang Zhou and Scott Hawken in a study of waste water treatment plants in Adelaide, Australia. Their analysis identified areas where the systems and coastal proximity lead to several vulnerabilities, including sea level rise (Figures 3.8 and 3.9) (Hawken et al., 2022). Modelling scenarios and alternatives (SSP 1-1.9 IPCC scenario), four land use options were devised by Zhou & Hawken (2023) to consider storm surge impact on infrastructure. The particular scenario aspect of geodesign enables various futures to be articulated from two important values from information sciences and through human values articulated by stakeholders and non-professionals, which in combination encourage convergence on outcomes through rationalised decision-making for resilience (Pettit et al., 2019).

GEODESIGN AS A FRAMEWORK FOR WORKING WITH UDTs

The geodesign framework, whilst procedural, can allow a number of steps to be rearranged (Yang & Delparte, 2022). Participants

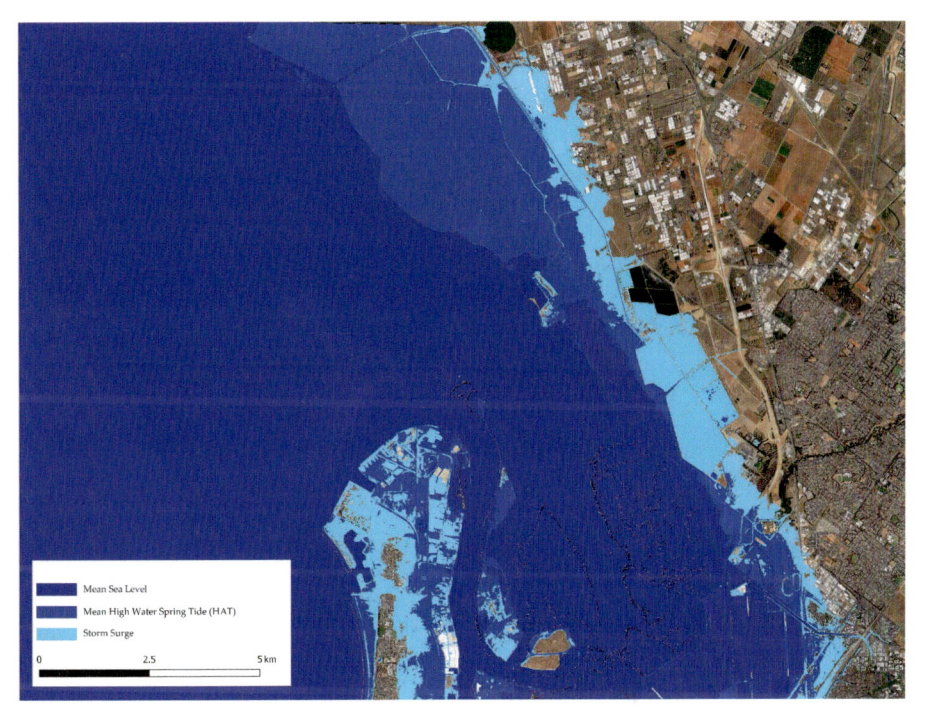

Figure 3.8
Kaihang Zhou and Scott Hawken, 2023. "Bolivar Water Treatment Plant vulnerability to flooding by 2100 under the High Predicted Inundation Scenario. The light blue indicates a storm surge. The mid-blue indicates high spring tide extent and deep blue indicates the mean sea level" (Zhou & Hawken, 2023, p. 16).

benefit from being involved in the process. It is essential to state that the framework is not without criticism. Richard Stiles criticises the hypothesis that a centralised encompassing system, such as the one Steinitz presents, can incorporate a wide variety of environmental disciplines around a unified theory of landscape when, at the same time, such professions are in flux or highly hybridised (1994, pp. 141–143). The relationships of actors in the process are also important. Actor relationships are also critical in the framework, especially concerning facilitators' power relationships in terms of directing decisions as seen in similar cases of participatory planning and the work of Lawrence Halprin (Hirsch, 2014). Additional issues arise from more-than-human perspectives in environmental planning, which the geodesign process needs to accommodate (Metzger in Davoudi et al., 2019). Further criticism revolves around the scales for the implementation of the geodesign process. However, recent publications and the IGC brief have sought to address global challenges (Steinitz et al. in Droege, 2023; Steinitz et al., 2023). Other criticisms of the variability and process of national planning systems in different countries have led to some criticism of the broad suitability of geodesign. Critically, in this book, there is also a technological relationship that is both enabling and reliant in terms of progression and reporting using the software. As Mathew Wilson notes,

> **geodesign is both an idea and an investment that culminates in software. As software is ultimately conditioned by its potential markets, the central problematic for a more critical geodesign becomes how to proceed (and we must) while negotiating this conditioning.**
> (Wilson, 2015, p. 232)

Geodesign can be intrinsically part of these technological aspects as part of remote sensing, reality capture, 3D geospatial, BIM, AI urban analytics, Internet of Things (IoT), data dashboards, cloud computing, and many others alongside generative AI, large language models, also mean the role of geodesign is under consistent refinement in relation to the use of these processes, objects, and tools (Ervin, 2016; Tulloch, 2016; Afrooz et al., 2018; Dawkins et al., 2018). Indeed, commercial and open-source approaches can be utilised in hybrid ways. The geodesign framework and the six-step process are shown here in Figure 3.9. The methods of investigation are also open and can be chosen to suit the study area, resource capabilities, and brief timeframe. Indeed, the framework can be altered and adapted for other environmental disciplines. The following descriptions of the model processes are adapted from and have additional mappings to applications of UDTs by the author (Perkl, 2016) (Figure 3.10).

Figure 3.9
Kaihang Zhou and Scott Hawken, 2023. Mapping of four scenarios for land use of urban and natural systems. (a) The business-as-usual approach includes (b) a protection scenario, (c) a retreat scenario, and (d) an accommodation scenario.

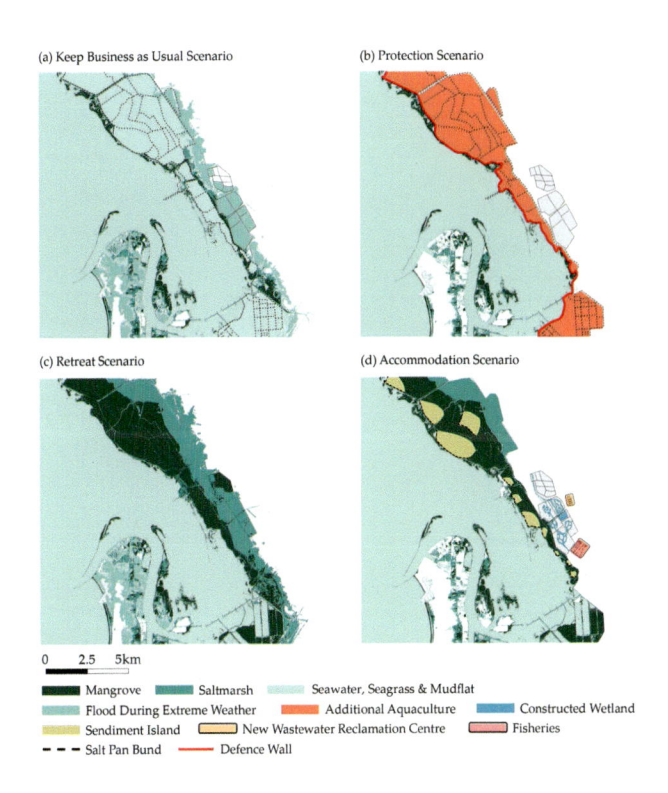

Assessment

Representation Model – Amalgamation of inventoried data that represents the study area and objectives. Organised and visualised to show all components and feed models.

Process Model – Exploring the relationships and interactions between data. This could include suitability, permeability, dynamic surfaces, and morphology, amongst other attributes.

Evaluation Models – Represents the modelled landscape and results as indicators and outputs.

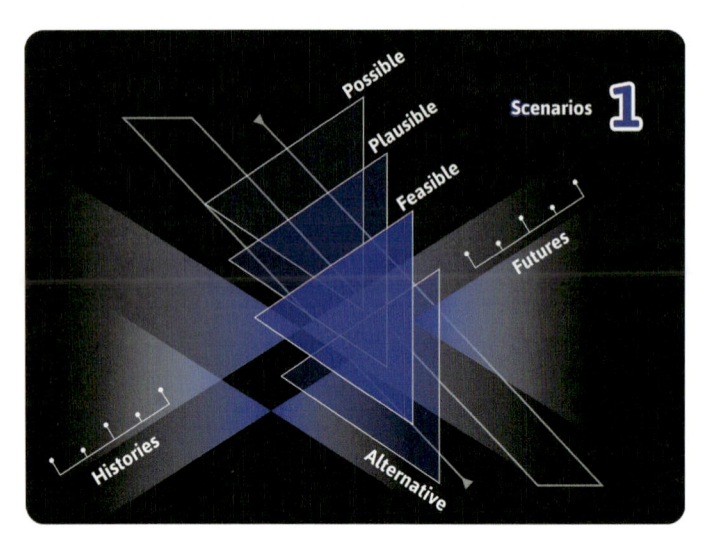

Figure 3.10
Authour, Scenarios card for mapping Plausible, Preferable and improbable Futures for geodesign, adapted from Voros (2003). Framework process UX Geodesign: A Collaborative Framework. Available at Figma, https://www.figma.com/community/file/1253768043463157544/Geodesign-a-Collaborative-Framework.

Intervention

Change Model – These are altered versions of the process and evaluation models that explore changes. This could be alterations to habitat, connectivity, or urban growth.

Impact Model – This process evaluates and quantifies the effect of the change models and the metrics used for evaluation. This could be a change detection process between the original representation and the change model.

Decision Models – Decision and Evaluation models rely upon stakeholder input to establish benchmarks to evaluate performance and make decisions. Additional iterations may be required if the designs are not satisfactory in feedback.

The Steinitz geodesign framework applied to UDTs could feature the following qualities, stages, and iterations to which I have defined as people, models, and platforms. These three overlapping areas are loosely described and would constitute the stakeholders, model approaches, and platforms for decisions and engagement in a UDT.

People

Representation Model – Concerning an Urban Digital Twin, this involves the specification and standards, the geographic study area being established, and baseline information.

Process Model – This stage in relation to a UDT concerns the modelling choices, spatiotemporal dynamics, resolutions, collection methods, storage and security and the dynamics between them.

Models

Evaluation Model – Represents the established system of system approach, containing connected maps, models, and sensors that 'twin' the geographic area in question and establish key indicators.

Change Model – Runs the simulative and predictive scenarios based on the evaluation models and key indicators based on historical or near-real-time data and constitutes the platform of the UDT.

Platforms

Impact Model – Displays modelled results usually via dashboard approaches on the established platform to quantify the impact of change models on the study area.

Decision Model – is the critical socio-technical interface between stakeholders and a UDT for collaborative decisions based on the impact models and scenarios.

Using the geodesign process for systematic procedural steps through study areas lends itself to the variety of UDTs and models. CIMs are situated between BIM and GIS, and IoT, and as Chapter 2 argues, are pre-cursors to UDTs (Cureton & Hartley, 2023). This crossover and geodesign approach for UDTs were explored in the UK in a 'CIM Forum,' which scoped remote sensing, 3D geospatial, CIMs, and UDT opportunities across the UK from a wide range of stakeholders from UK Government departments, local planning authorities, commercial companies, and academic researchers using some geodesign processes such as data needs and future scenarios (Figure 3.11). The full framework was not implemented in this case, but various scenarios were generated and evaluated, creating some early testing of the geodesign and UDT hypothesis.

The use of geodesign change models requires additional discussion and unpacking. Change models can be approached from nine different routes: anticipatory, participatory, sequential,

Figure 3.11
Authour, City Information Modelling Workshop (CIM Forum), Lancaster University, UK, September 12, 2023.

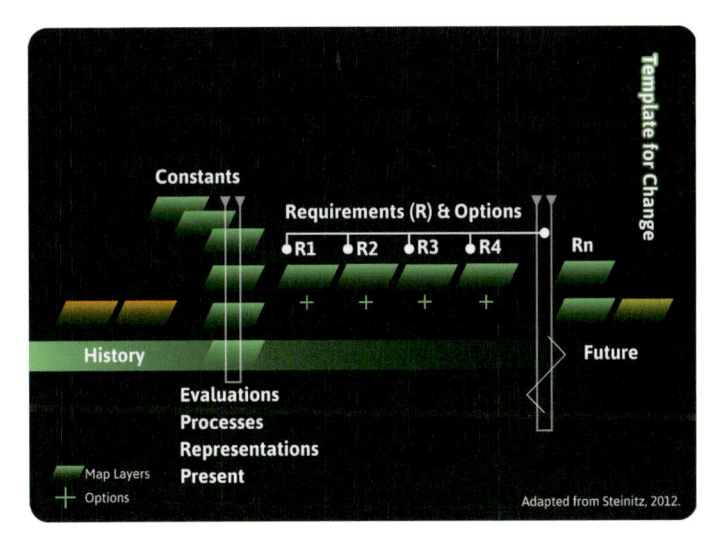

Figure 3.12
Change Model Template. Adapted from Steinitz (2012). Framework process UX Geodesign: A Collaborative Framework. Available at Figma, https://www.figma.com/community/file/1253768043463157544/Geodesign-a-Collaborative-Framework.

constraining, combinatorial, rule-based, optimised, agent-based, and mixed (Steinitz, 2012, pp. 56–59) and Steinitz presents a template in which the various routes can be mapped (Figure 3.12). Bridging the theorisation of the framework and practice, it has been applied and tested in a variety of contexts in three case studies for territorial resilience via the use of GIS cloud platforms and procedural modelling for the Metropolitan City of Cagliari and sustainable tourism development in the Oristano Gulf, Sardinia, Italy, and Urban development of the Meridia Neighbourhood in Nice, France (Caglioni & Campagna, 2021). The geographies and countries to which the geodesign process has been applied are highly varied, from global cities to rural hinterlands and coastal zones (Nyerges et al., 2016; McNally et al., 2021). A complete systematic literature review of geodesign took place by Ripan Debnath et al., of which 487 cases were identified and clustered, 75 cases were related to resilience, 41 involved computational aspects, and 18 to collaborative possibilities of the framework. The study concluded that (n = 75) real untapped potential existed in future work with geodesign around planning resilience (Debnath et al., 2022). The wide variety of case studies and varying contexts, as well as the longevity of the framework, naturally align with the emergent position of UDTs and provide a critical socio-technical framework for operation.

As geodesign has varying technological relationships, there is a natural synergy in UDTs, but that also does not mean the extent of studies should be urban in nature, as rural development and agriculture also play a role, as referenced for Łódź region, Poland (Wójcik et al., 2021) and Idaho, USA (Cronan et al., 2023). In addition, the translation to critical areas, such as the expansion of industrial zones, evaluation, and scenario planning, has been conducted by Lisha He and Jian Zhang and Yuhan Huo, (Figure 3.13), indicating the framework's flexibility. Additional research gaps have been identified by Xinyue Ye et al. in geodesign processes and agent-based simulations of the cascading impacts of climate hazards (Ye et al., 2023, p. 195). The critical issue in implementing geodesign is the participatory process and the bringing together of enabling communities, which Randolph Hester has advocated in careful consideration (2010). In the case of Houseal Lavigne, map.social is a GIS platform for collective mapping for user-generated content and participatory GIS to create stories of places via a design charette (Figure 3.14).

Tessa Eikelboom and Ron Janssen created four geodesign tools with real-time analysis and interaction via a visualisation table with two stakeholder groups to explore scenarios for climate mitigation for a peat meadow in the northern part of the Netherlands, resulting in the authors stating that "careful selection of methods and tools supports the development of adaptation plans and rationality can be used to choose between different geodesign tools" (Eikelboom & Janssen, 2017, p. 264) (Figure 3.15). This particular experiment would be one aspect of the relationship between geodesign and new near real-time data attributes that can be found in a UDT. Finally,

CHAPTER 3
GEODESIGN AND URBAN DIGITAL TWINS

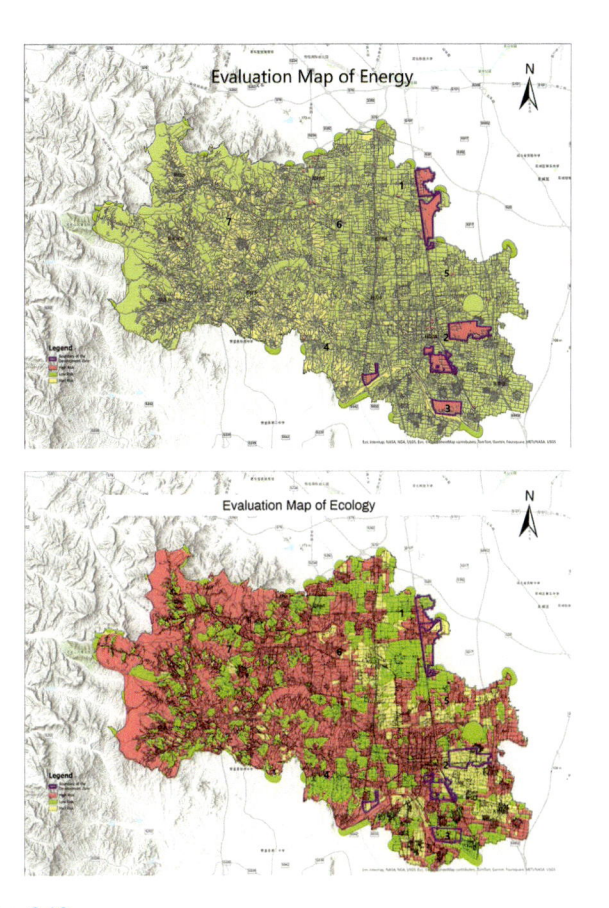

Figure 3.13
Lisha He, PhD Candidate Lancaster University, Jian Zhang, Yuhan Huo, Hebei Provincial Planning Institute (HPPI), Geodesign Evaluation Maps, Ecology, and Energy (two of ten), Yuanshi industrial zone, Yuanshi County, Hebei, China, 2024. In this case, during the evaluation geodesign phase, He and Zhang mapped high-risk, medium, and low-risk areas. For example, in the Energy Evaluation map, "High vulnerability areas are characterised by high total and per capita energy consumption, indicating heavy dependence and low productivity per unit of energy. They also exhibit high energy intensity, requiring large amounts of energy per unit of economic output, an unfavourable energy mix with heavy reliance on a single or expensive energy source, and high energy costs that can affect economic stability. Medium Vulnerability Area: Likely to exhibit medium levels of the above factors, showing a balanced treatment of energy use. Areas of low energy vulnerability should ideally have low overall and per capita energy consumption, indicating lower dependence and higher efficiency. They should also exhibit low energy intensity, a diversified and cost-effective energy mix, and low energy costs, collectively ensuring greater economic resilience and efficiency in converting energy into economic output" (He, 2024).

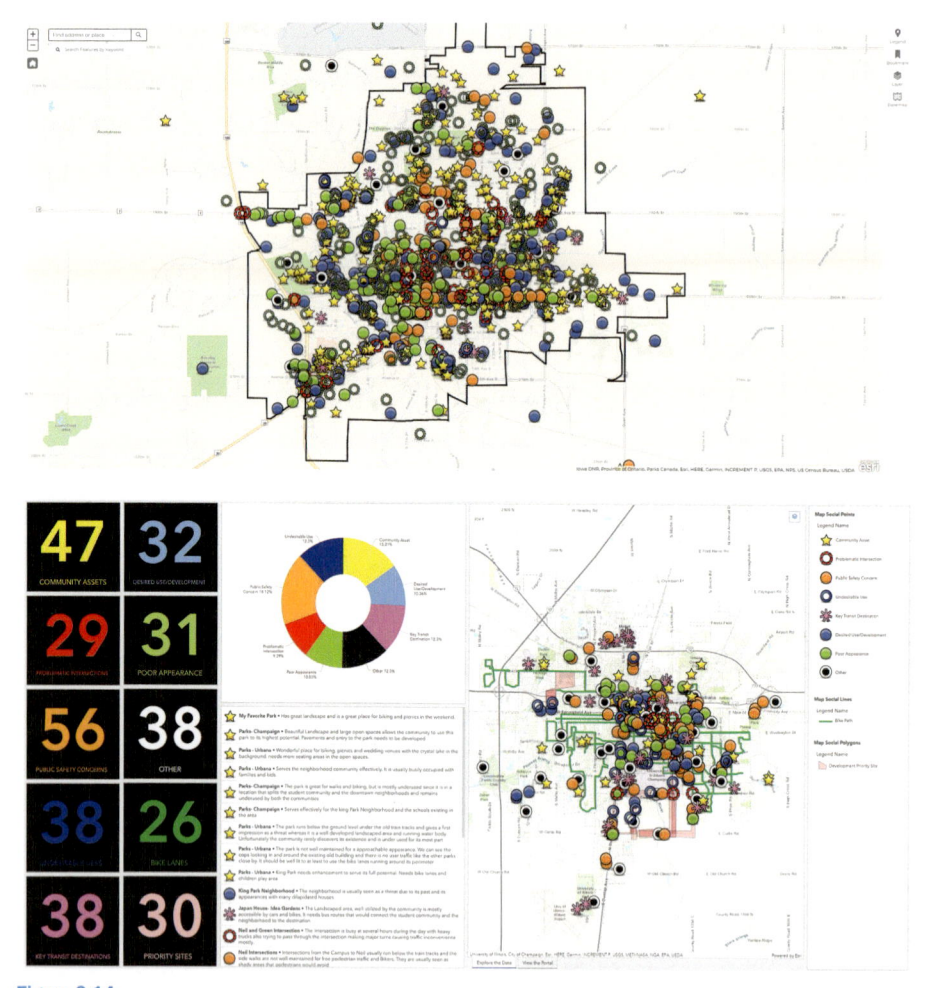

Figure 3.14
Map.Social. A cloud-based mapping platform for collecting community input. The app allows users to add customized points to the map to express a concern or highlight an issue. Users can then up or down-vote items so that important issues are highlighted.

supporting the hypothesis, a study by Srivastava et al. (2022) utilised 3D GIS scenes and a CAVE environment for future geodesign scenarios at the Mooloolaba Spit area in the south-east Queensland region of Australia for an immersive experience into the implications of planning decisions (Figure 3.16).

CHAPTER 3
GEODESIGN AND URBAN DIGITAL TWINS

Figure 3.15
Tessa Eikelboom and Ron Janssen, 2017, Collaborative use of geodesign tools to support decision-making on adaptation to climate change. "Examples of measures applied during the experiment with the original situation on the left and the changes on the right for each tool" (Eikelboom & Janssen, 2017, p. 260).

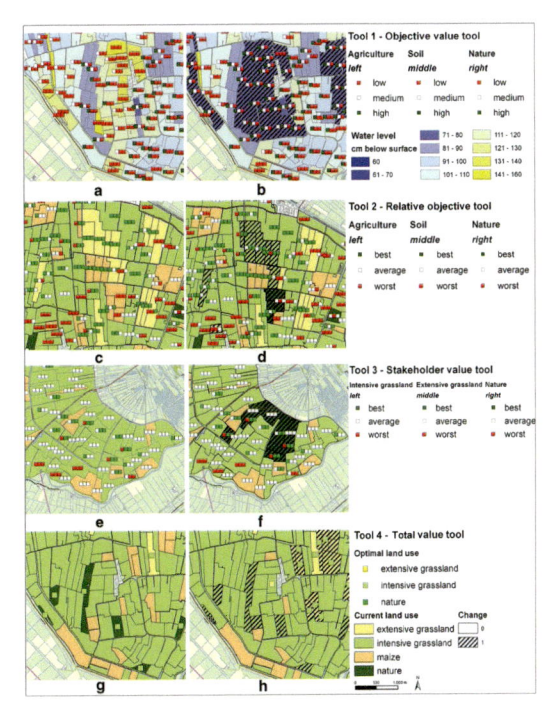

Figure 3.16
Sanjeev Kumar Srivastava, Gary Scott, and Johanna Rosier, "A collection of scenarios displayed across a variety of visualisation media. (a) Recent scenario, (b) past scenario, (c) recent scenario with 3d models, (d) past scenario with 3d models, (e) alternate scenario in CityEngine web browser, (f) alternate scenario on a mobile device and (g) alternate scenario in a real 3d (CAVE2) immersive environment" (Srivastava et al., 2021, p. 30).

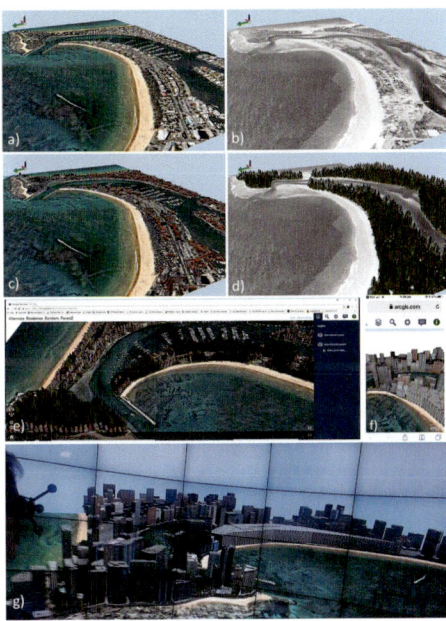

159

SUMMARY

The many possibilities offered by various models align to certain degrees of a replica of the physical world that have been discussed here, offering information-rich technological support for the geodesign framework but remains a working hypothesis for suitability. One of the critical research gaps is identifying suitable methods for geodesign and UDTs, which is discussed in Chapter 4. UDTs and the geodesign process are emergent but high-potential research areas that seek approaches to address complexity. Methodological choices have become quite important, and the UN-Habitat Block-by-Block playbook methodology launched in 2012 is a useful case study in participation through the use of Mojang's Minecraft, community meetings, presentations, and site visits in six phases similar to the geodesign framework: (1) plan, (2) design, (3) develop, (4) operate, (5) monitor, and (6) evaluate. Several participatory cases were featured, though in Hanoi, Vietnam, a project sought to engage 120 schoolchildren to make streets safer for young girls who commute to school, resulting in upgrades to underpasses and other spaces (Figure 3.17). Block-by-Block released an open playbook and has delivered projects across the world. Digital games are particularly pertinent in providing low-cost modelling tools, dialogue, and gamification of urban environments.

The geodesign framework builds upon various theories of systems thinking and has a reciprocal relationship with maps and GIS. Technology is not emphasised, but the dialogue, deliberation, and consensus-building possible through the process are conducive towards UDTs. Commercial advocacy of geodesign by ESRI, along with the geodesign hub and IGC, continues to implement geodesign projects. Criticism has been focused on the technological relationship, the role of actors in the geodesign process, scales of application, and the rationalised procedural nature of the framework. However, the wide variety of geographic cases and the adaptability of the framework support the continuing longevity of geodesign at the same time. Geodesign is not a panacea for delivering consensus and negotiation of complex environments (Flint Ashery & Steinitz, 2022).

Figure 3.17
U.N. Habitat, Katla Studios, and Block-by-Block. Hanoi, Vietnam, Plan International, North Thang Long Economic & Technical College, 2017. https://www.blockbyblock.org/.

Indeed, the selected outcomes from decision models may not always be appropriate for tangible futures and be best fits for climate adaptation. Geodesign iterations and 'model' progressions also require careful facilitation and design. However, emergent UDTs are to come to the fore, offer tangible social benefits, and enrich decision-making for urban management. In that case, we do not require novelty, new frameworks, or entirely new formations but a combination of an established framework to negotiate the future.

The multidisciplinary convergence of environmental science to deal with systems and complexity and extreme climate challenges while simultaneously engaging multiple stakeholders requires a framework to accommodate such a demanding ecosystem. Arguably, Utilising the latest modelling procedures and technological systems, an established geodesign framework can assess, select, and design the most optimal futures to address our most pressing climate challenges.

CASE STUDY: SHANNON MCELVANEY GLOBAL TECHNOLOGY LEAD, GEODESIGN – ADVANCE PLANNING GROUP, JACOBS

The Meridian Water redevelopment project is a major £6bn ($7.8bn), 20-year London regeneration program led by Enfield Council, bringing 10,000 homes and 6,000 jobs to Enfield, north London. Alongside beautiful homes and world-class public spaces and community facilities, the 200-acre development has its own brand-new railway station, linking commuters to the region.

Jacobs was responsible for creating the data visualisation platform and KPI-driven geodesign processes required to synthesise data coming from multi-sources into a common framework to support decision-making. The result was a set of live, interactive 2D and 3D dashboards and models to help optimise the master planning process to ensure that it met the sustainability and resilience goals set forward by the council, including: (1) a vibrant mix-used development with animated streets and a park at your doorstep; (2) access to multi-modal transit and pedestrian-friendly streets; and (3) a vibrant setting for shops, restaurants, and businesses with ample affordable housing.

To ensure everyone had "a park at their doorstep," that simple project brief was translated into several metrics and criteria. For example, 30% of the site was earmarked as open space and a requirement that all parks would be no more than 400 m from any doorstep was set (Figure 3.18). Spatial analytics insured that both criteria could be met and/or exceeded, including the criteria that every child would have 10 m^2 of play space.

The second part of the brief was to support multi-modal transport and equal access to rail, essentially to create a walkable, bikeable,

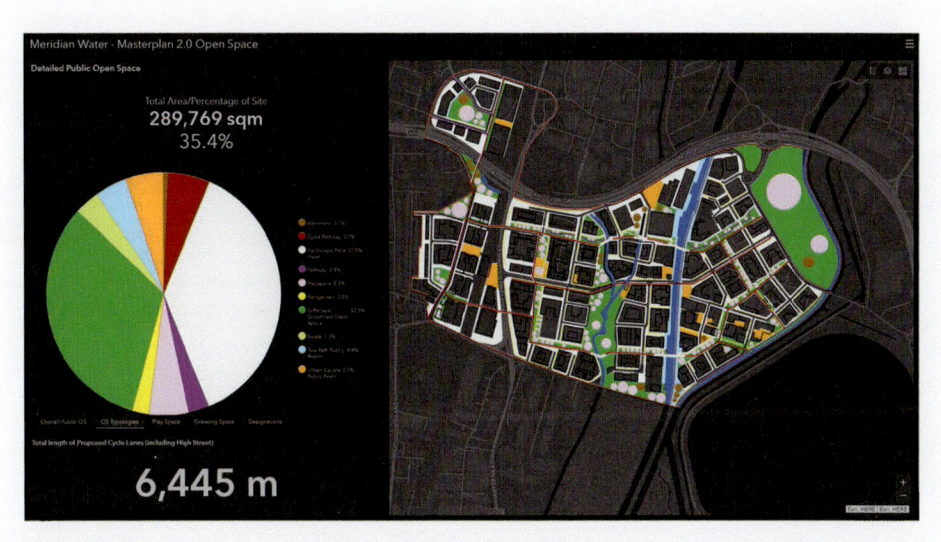

CHAPTER 3
GEODESIGN AND URBAN DIGITAL TWINS

Figure 3.18
A 2D interactive thematic map and dashboard detailing ten open space typologies as well as well as the total area of coverage. Multiple scenarios and views could be selected using map tabs, a great way of giving stakeholders the opportunity to investigate what was important to them.

Figure 3.19
Esri's ArcGIS CityEngine and the Complete Streets Tool were used to generate hypothetical streetscape scenarios with different modal mixes of bus and bike lanes as well as sidewalks. Trees and different amenities were added to give stakeholders a way of visualising future scenarios before making a decision.

163

transit-oriented development. Again, this was translated into 400-m walking distances from every doorstep to shops, restaurants, and the rail station. And street typologies were created and used to generate different streetscapes for modal comparison and selection (Figure 3.19).

https://www.jacobs.com/newsroom/news/meridian-water-sustainable-and-resilient-design

REFERENCES

Afrooz, A., Ballal, H., & Pettit, C. (2018). Implementing augmented reality sandbox in geodesign: A future. *ISPRS Annals of the Photogrammetry, Remote Sensing and Spatial Information Sciences*, 4, 5–12. https://doi.org/10.5194/isprs-annals-IV-4-5-2018.

Albert, C., Brillinger, M., Guerrero, P. et al. (2021). Planning nature-based solutions: Principles, steps, and insights. *Ambio*, 50, 1446–1461. https://doi.org/10.1007/s13280-020-01365-1.

Banathy, B.H. (1996). *Designing social systems in a changing world* (p. XV, 372). Springer US. ISBN 978-0-306-45251-2.

Batty, M. (2018). Digital twins. *Environment and Planning B: Urban Analytics and City Science*, 45(5), 817–820. https://doi.org/10.1177/2399808318796416.

Blair, G.S. (2021). Digital twins of the natural environment. *Patterns*, 2(10), 100359, ISSN 2666-3899. https://doi.org/10.1016/j.patter.2021.100359.

Bolton, A., Butler, L., Dabson, I., Enzer, M., Evans, M., Fenemore, T., Harradence, F. et al. (2018). *The Gemini principles*. Centre for Digital Built Britain. Accessed 10 Sep 2023. https://www.cdbb.cam.ac.uk/DFTG/GeminiPrinciples.

Caglioni, M., & Campagna, M. (2021). Geodesign for collaborative spatial planning: Three case studies at different scales. In E. Garbolino, & C. Voiron-Canicio (Eds.), *Ecosystem and territorial resilience* (pp. 323–345). Elsevier. ISBN 9780128182154. https://doi.org/10.1016/B978-0-12-818215-4.00012-2.

Caldarelli, G., Arcaute, E., Barthelemy, M. et al. (2023). The role of complexity for digital twins of cities. *Nature Computational Science*, 3, 374–381. https://doi.org/10.1038/s43588-023-00431-4.

Chrisman, N. (2006). *Charting the unknown: How computer mapping at Harvard became GIS*. ESRI Press.

Cronan, D., Trammell, E.J., & Kliskey, A. (2023). From uncertainties to solutions: A scenario-based framework for an agriculture protection zone in Magic Valley Idaho. *Land*, 12, 862. https://doi.org/10.3390/land12040862.

Cureton, P., & Hartley, E. (2023). City Information Models (CIMs) as precursors for Urban Digital Twins (UDTs): A case study of Lancaster. *Frontiers in Built Environment*, 9, 1048510. https://doi.org/10.3389/fbuil.2023.1048510.

Cureton, P., & He, L. (2023). *Geodesign: A collaborative framework*. Figma Open. https://www.figma.com/communityfile/1253768043463157544/Geodesign-a-Collaborative-Framework.

Dawkins, O., Dennett, A., & Hudson-Smith, A.P. (2018). Living with a digital twin: Operational management and engagement using IoT and mixed realities at UCL's here east campus on the Queen Elizabeth Olympic Park. In *Proceedings of the 26th annual Giscience and remote sensing*. GIS Research UK (GISRUK).

Debnath, R., Pettit, C., & Leao, S.Z. (2022). Geodesign approaches to city resilience planning: A systematic review. *Sustainability*, 14, 938. https://doi.org/10.3390/su14020938.

Eikelboom, T., & Janssen, R. (2017). Collaborative use of geodesign tools to support decision-making on adaptation to climate change. *Mitigation and Adaptation Strategies for Global Change*, 22, 247–266. https://doi.org/10.1007/s11027-015-9633-4.

Ervin, S.M. (2016). Technology in geodesign. *Landscape and Urban Planning*, 156, 12–16, ISSN 0169-2046. https://doi.org/10.1016/j.landurbplan.2016.09.010.

Fisher, T., Orland, B., & Steinitz, C. (2020). *The international geodesign collaboration: Changing geography by design*. Esri Press.

Flaxman, M. (2010). Geodesign: Fundamental principles and routes forward. Talk at GeoDesign Summit.

Flint Ashery, S., & Steinitz, C. (2022). Issue-based complexity: Digitally supported negotiation in geodesign linking planning and implementation. *Sustainability*, 14, 9073. https://doi.org/10.3390/su14159073.

Foster, K. (2016). Geodesign parsed: Placing it within the rubric of recognized design theories. *Landscape and Urban Planning*, 156, 92–100, ISSN 0169-2046. https://doi.org/10.1016/j.landurbplan.2016.06.017.

Goodchild, M.F. (1992). Geographical information science. *International Journal of Geographical Information Systems*, 6(1), 31–45.

Gottwald, S., Brenner, J., Janssen, R. et al. (2021). Using Geodesign as a boundary management process for planning nature-based solutions in river landscapes. *Ambio*, 50, 1477–1496. https://doi.org/10.1007/s13280-020-01435-4.

Hawken, S., Zhou, K., Mosley, L., & Leyden, E. (2022). Scenario-based thinking to negotiate coastal squeeze of ecosystems: Green, blue, grey, and hybrid infrastructures for climate adaptation and resilience. In M. Jarchow, & A. Rastandeh (Eds.), *Creating resilient landscapes in an era of climate change: Global case studies and real-world solutions* (pp. 231–250). Routledge. https://doi.org/10.4324/9781003266440.

He, L., Geodesign Evaluation Maps, Interview with Paul Cureton (03/04/24).

Herrington, S. (2010). The nature of Ian McHarg's science. *Landscape Journal*, 29(1), 1–20. https://doi.org/10.3368/lj.29.1.1.

Hessler, J. (2015). Computing space III: Computing space 0: From hypersurfaces to algorithms: Saving early computer cartography at the library of congress, November 2015. Accessed 8 Dec 2023. https://blogs.loc.gov/maps/2015/12/father-of-gis/.

Hester, R. (2010). *Design for ecological democracy*. MIT Press.

Hirsch, A.B. (2014). *City choreographer: Lawrence Halprin in urban renewal America*. University of Minnesota Press.

Hollstein, L.M. (2019). Retrospective and reconsideration: The first 25 years of the Steinitz framework for land- scape architecture education and environmental design. *Landscape and Urban Planning*, 186, 56–66, ISSN 0169-2046. https://doi.org/10.1016/j.landurbplan.2019.02.020.

Hudson-Smith, A., & Batty, M. (2022). Ubiquitous geographic information in the emergent Metaverse. *Transactions in GIS*, 26, 1147–1157. https://doi.org/10.1111/tgis.12932.

IET. (2022). The Apollo Protocol: Unifying digital twins across sectors, Monday, 12 September 2022. https://www.theiet.org/impact-society/factfiles/built-environment-factfiles/the-apollo-protocol-unifying-digital-twins-across-sectors.

Kingston, R., Hill, K., & Poplin, A. (2020). Geodesign systems. In T. Fisher, B. Orland, & C. Steinitz (Eds.), *The international Geodesign collaboration: Changing geography by design*. Esri Press.

Klosterman, R.E. (1999). The what if? Collaborative planning support system. *Environment and Planning B: Planning and Design*, 26(3), 393–408. https://doi.org/10.1068/b260393.

Kunzmann, K.R. (1993). Geodesign: Chance oder Gefahr? In Bundesforschungsanstalt für Landeskunde und Raumordnung (Ed.), *Planungskartographie und. Geodesign* (Vol. 7, pp. 389–396). Informationen zur Raumentwicklung.

Lee, M.-C. (2016). Geodesign scenarios. *Landscape and Urban Planning*, 156, 9–11, ISSN 0169-2046. https://doi.org/10.1016/j.landurbplan.2016.11.009.

Li, Y., & Kim, Y. (2022). Analysis of effects of sponge city projects applying the geodesign framework. *Land*, 11, 455. https://doi.org/10.3390/land11040455.

Li, X., Feng, M., Ran, Y. et al. (2023). Big data in earth system science and progress towards a digital twin. *Nature Reviews Earth and Environment*, 4, 319–332. https://doi.org/10.1038/s43017-023-00409-w.

Longley, P., Goodchild, M., Maguire, D.J., & Rhind, D.W. (2001). *Geographic information systems and science*. Wiley.

Machl, T., Donaubauer, A., & Kolbe, T.H. (2019). Planning agricultural core road networks based on a digital twin of the cultivated landscape. *Journal of Digital Landscape Architecture*, 4, 316–327.

McElvaney, S. (2012). *Geodesign: Case studies in regional and urban planning*. ESRI Press.

McNally, B., Power, P., & Foley, K. (2021). 'Going digital' – Lessons for future coastal community engagement and climate change adaptation. *Ocean & Coastal Management*, 208, 105629, ISSN 0964-5691. https://doi.org/10.1016/j.ocecoaman.2021.105629.

Metzger, J. (2019). A more-than-human approach to environmental planning. In S. Davoudi, R. Cowell, I. White, & H. Blanco (Eds.), *The Routledge companion to environmental planning* (1st ed.). Routledge. https://doi.org/10.4324/9781315179780.

Miller, W.R. (2012). *Introducing geodesign: The concept*. Esri Press.

Minner, J. (2017). Geodesign, resilience and the future of former mega-event sites. In S. Geertman, A. Allan, C. Pettit, & J. Stillwell (Eds.), *Planning support science for smarter urban futures* (CUPUM 2017. Lecture notes in geoinformation and cartography). Springer. https://doi.org/10.1007/978-3-319-57819-4_8.

Nochta, T. et al. (2019). The local governance of digital technology – Implications for the city-scale digital twin. *CDBB*. https://doi.org/10.17863/CAM.43321.

Nochta, T., Wan, L., Schooling, J.M., & Parlikad, A.K. (2021). A socio-technical perspective on urban analytics: The case of city-scale digital twins. *Journal of Urban Technology*, 28(1–2), 263–287. https://doi.org/10.1080/10630732.2020.1798177.

Nyerges, T., Ballal H., Steinitz, C., Canfield, T., Roderick, M., Ritzman, J., Thanatemaneerat, & W. (2016). Geodesign dynamics for sustainable urban watershed development. *Sustainable Cities and Society*, 25, 13–24, ISSN 2210-6707. https://doi.org/ 10.1016/j.scs.2016.04.016.

Orland, B. (2015). Commentary: Persuasive new worlds: Virtual technologies and community decision-making. *Landscape and Urban Planning*, 142, 132–135, ISSN 0169-2046. https://doi.org/10.1016/j.landurbplan.2015.08.010.

Orland, B., & Murtha, T. (2015). Research article: Show me: Engaging citizens in planning for shale gas development. *Environmental Practice*, 17(4), 245–255. https://doi.org/10.1017/S1466046615000290.

Orland, B. (2016). Geodesign to tame wicked problems. *Journal of Digital Landscape Architecture*, 1, 187–197. https://doi.org/10.14627/537612022.

Paradis, T., Treml, M., & Manone, M. (2013). Geodesign meets curriculum design: Integrating geodesign approaches into undergraduate programs. *Journal of Urbanism*, 6(3), 274–301. https://doi.org/10.1080/17549175.2013.788054.

Perkl, R.M. (2016). Geodesigning landscape linkages: Coupling GIS with wildlife corridor design in conservation planning. *Landscape and Urban Planning*, 156, 44–58, ISSN 0169-2046. https://doi.org/10.1016/j.landurbplan.2016.05.016.

Pettit, C.J., Hawken, S., Ticzon, C., Leao, S.Z., Afrooz, A.E., Lieske, S.N., Canfield, T., Ballal, H., & Steinitz, C. (2019). Breaking down the silos through geodesign – Envisioning Sydney's urban future. *Environment and Planning B: Urban Analytics and City Science*, 46(8), 1387–1404. https://doi.org/10.1177/2399808318812887.

Rittel, H.W.J., & Webber, M.M. (1973). Dilemmas in a general theory of planning. *Policy Sciences*, 42(4), 155–169. https://doi.org/10.1007/BF01405730.

Robertson, J.C. (1967). The Symap programme for computer mapping. *The Cartographic Journal*, 4(2), 108–113. https://doi.org/10.1179/caj.1967.4.2.108.

Schroth, O. (2023). Geodesign as online teaching method – Lessons from a multiple case study. *Journal of Digital Landscape Architecture*, 2023(8), 598–6072023.

Shakeri, M., Kingston, R., & Pinto, N. (2016). Game science or games and science? Towards an epistemological understanding of use of games in scientific fields. In T. Marsh, M. Ma, M.F. Oliveira, J.B. Hauge, & S. Göbel (Eds.), *Serious games: Second joint international conference*, JCSG 2016, Brisbane, QLD, Australia, September 26–27, 2016, proceedings (Lecture notes in computer science) (Vol. 9894, pp. 163–168). Springer Nature. https://doi.org/10.1007/978-3-319-45841-0_15.

Shearer, A.W. (2022). Expanding the use of scenarios in geodesign: Engaging uncertainty of the anthropocene. *JoDLA – Journal of Digital Landscape Architecture*, 7. https://doi.org/10.14627/537724046.

Shepard, D.S. (1984). Computer mapping: The SYMAP interpolation algorithm. In G.L. Gaile, & C.J. Willmott (Eds.), *Spatial statistics and models (Theory and decision library)* (Vol. 40). Springer. https://doi.org/10.1007/978-94-017-3048-8_7.

Somma, M., Campagna, M., Canfield, T., Cerreta, M., Poli, G., & Steinitz, C. (2022). Collaborative and sustainable strategies through geodesign: The case study of Bacoli. In O. Gervasi, B. Murgante, S. Misra, A.M.A.C. Rocha, & C. Garau (Eds.), *Computational science and its applications – ICCSA 2022 workshops*. ICCSA 2022 (Lecture notes in computer science) (Vol. 13379). Springer. https://doi.org/10.1007/978-3-031-10545-6_15.

Srivastava, S.K., Scott, G., & Rosier, J. (2021). Use of geodesign tools for visualisation of scenarios for an ecologically sensitive area at a local scale. *Environment and Planning B: Urban Analytics and City Science.* https://doi.org/10.1177/2399808321991538.

Srivastava, S.K., Scott, G., & Rosier, J. (2022). Use of geodesign tools for visualisation of scenarios for an ecologically sensitive area at a local scale. *Environment and Planning B: Urban Analytics and City Science,* 49(1), 23–40. https://doi.org/10.1177/2399808321991538.

Steinitz, C., & Rogers, P.P. (1970). *A systems analysis model of urbanization and change an experiment in interdisciplinary education.* MIT Press.

Steinitz, C. (1990). A framework for theory applicable to the education of landscape architects (and other environmental design professionals). *Landscape Journal,* 9(2), 136–143.

Steinitz, C., Arias, H., Bassett, S., & Flaxman, M. (2002). *Alternative futures for changing landscapes: The upper San Pedro River basin in Arizona and Sonora.* Island Press.

Steinitz, C. (2012). *A framework for geodesign: Changing geography by design.* ESRI Press.

Steinitz, C. (2013). Beginnings of geodesign: A personal historical perspective. Accessed 9 Mar 2022. https://www.esri.com/about/newsroom/arcnews/beginnings-of-geodesign-a-personal-historical-perspective/.

Steinitz, C., Orland, B., Fisher, T., & Campagna, M. (2023). Geodesign to address global change. In P. Droege (Ed.), *Intelligent environments* (2nd ed., pp. 193–242). North-Holland. ISBN 9780128202470. https://doi.org/10.1016/B978-0-12-820247-0.00016-3.

Stiles, R. (1994). Landscape theory: A missing link between landscape planning and landscape design? *Landscape and Urban Planning,* 30(3), 139–149, ISSN 0169-2046. https://doi.org/10.1016/0169-2046(94)90053-1.

Tran, D.X., Pearson, D., Palmer, A., Dominati, E.J., Gray, D., & Lowry, J. (2023). Integrating ecosystem services with geodesign to create multifunctional agricultural landscapes: A case study of a New Zealand hill country farm. *Ecological Indicators,* 146, 109762, ISSN 1470-160X. https://doi.org/10.1016/j.ecolind.2022.109762.

Tulloch, D. (2016). Relinquishing a bit of control: Questions about the computer's role in geodesign. *Landscape and Urban Planning,* 156, 17–19, ISSN 0169-2046. https://doi.org/10.1016/j.landurbplan.2016.09.007.

Virdee, M., & Hughes, M. (2022). *Why did nobody see it coming? How scenarios can help us prepare for the future in an uncertain world.* RAND Europe and the RAND Europe Centre for Futures and Foresight Studies, January 28, 2022. https://www.rand.org/pubs/commentary/2022/01/why-did-nobody-see-it-coming-how-scenarios-can-help.html.

Voros, J. (2003). A generic foresight process framework. *Foresight*, 5(3), 10–21. https://doi.org/10.1108/14636680310698379.

Wilson, M.W. (2015). On the criticality of mapping practices: Geodesign as critical GIS? *Landscape and Urban Planning*, 142, 226–234, ISSN 0169-2046. https://doi.org/10.1016/j.landurbplan.2013.12.017.

Wójcik, M., Dmochowska-Dudek, K., & Tobiasz-Lis, P. (2021). Boosting the potential for GeoDesign: Digitalisation of the system of spatial planning as a trigger for smart rural development. *Energies*, 14, 3895. https://doi.org/10.3390/en14133895.

Wu, C.-L., & Chiang, Y.-C. (2018). A geodesign framework procedure for developing flood resilient city. *Habitat International*, 75, 78–89, ISSN 0197-3975. https://doi.org/10.1016/j.habitatint.2018.04.009.

Yang, X., & Delparte, D. (2022). A procedural modelling approach for ecosystem services and geodesign visualization in old town Pocatello, Idaho. *Land*, 11, 1228. https://doi.org/10.3390/land11081228.

Ye, X., Du, J., Han, Y., Newman, G., Retchless, D., Zou, L., Ham, Y., & Cai, Z. (2023). Developing human centered urban digital twins for community infrastructure resilience: A research agenda. *Journal of Planning Literature*, 38(2), 187–199. https://doi.org/10.1177/08854122221137861.

Zhou, K., & Hawken, S. (2023). Climate-related sea level rise and coastal wastewater treatment infrastructure futures: Landscape planning scenarios for negotiating risks and opportunities in Australian urban areas. *Sustainability*, 15(11), 8977. https://doi.org/10.3390/su15118977.

Zlotin, B., & Zusman, A. (2013). Inventive problem solving (TRIZ), theory. In E.G. Carayannis (Ed.), *Encyclopaedia of creativity, invention, innovation, and entrepreneurship*. Springer. https://doi.org/10.1007/978-1-4614-3858-8_36.

CHAPTER 4

GEODESIGN METHODS FOR URBAN DIGITAL TWINS

INTRODUCTION

Urban digital twins and federated systems have a high likelihood of becoming ubiquitous being applied in specific use cases, such as energy, transport, or planning development from a smart city perspective. However, the form of development is critically important, and Chapter 3 has argued for geodesign as a collaborative framework for implementation. As we have seen in Chapter 2, UDT ambitions emerge from CIM approaches and the use of 3D GIS for baseline information, urban dashboards for urban indicators for governance, and XR interactions for participation, though these have often been conducted as standalone experiments without integration. However, this implementation will have significant challenges in terms of approaches, complexity science, and execution. Research teams face several challenges, including the scale of the UDT, choice of modelling approaches, variety of applications, scales of operation, duration of the UDT, and interoperability. Interoperability also depends on the internal system architecture and how the UDT communicates with other UDTs. UDTs will contain multiple independent systems which form as a whole,

> **A system of systems approach applies systems thinking to the built and natural environment.**
> **A system is a connected collection of interrelated and interdependent parts; a complex whole that may be more than the sum of its parts. It is influenced by its environment, defined by its structure and purpose, and expressed through its function.**
> (Council & Lamb, 2022, p. 8)

This chapter builds upon the frameworks for geodesign, which naturally aligns with systems thinking through its procedural approach.

Carl Steinitz has noted that there are no prescribed methods for geodesign, though he notes a number of computational precedents, such as agent-based modelling (ABM) and rule-based change models (Steinitz, 2012, pp. 140–141). There are also several parallel approaches to other design and systems theories, which have been extensively discussed by Kelleann Foster (2016) that can also be mapped. However, there is a specific research gap in mapping the geodesign framework to the systems-based approach of UDTs, which this chapter addresses as a contribution to knowledge. The methods discussed here incorporate recent tools and modes for participation and collaboration using advanced geographic systems in the built environment and futuring methods. Urban planning is often faced with 'Wicked Problems': "the definition of a wicked problem is the problem itself" (Rittel & Webber, 1973, p. 161). Futuring as a methodological approach can also offer a meaningful exchange for wicked problems in urban planning (Hoffman et al., 2021, p. 2).

The novelty of this book lies in the connection between an established framework for decision-making for dealing with complexity and emerging technology embodied by systems of systems approaches. For geodesign to work as a 'thinking' framework for analysis and decision-making for UDTs, the iterative aspect of the chosen methods is very important as this prepares for two important steps in geodesign. The first is the collaborative and editorial role that collective iterations of the six stages of geodesign have: representation, process, evaluation, change, and impact. Secondly, the future-facing aspect for simulation and scenario planning that the chosen methods embody. These methods are not traditionally part of geodesign, but they are associated with the *assessment* and *intervention* phases, which are two of the six stages of the geodesign framework (Figure 4.1).

This chapter discusses a range of overlapping methods for UDT creation based on the geodesign framework and provides an overview

Figure 4.1
Authour, Eleven Geodesign Method Cards, 2024.

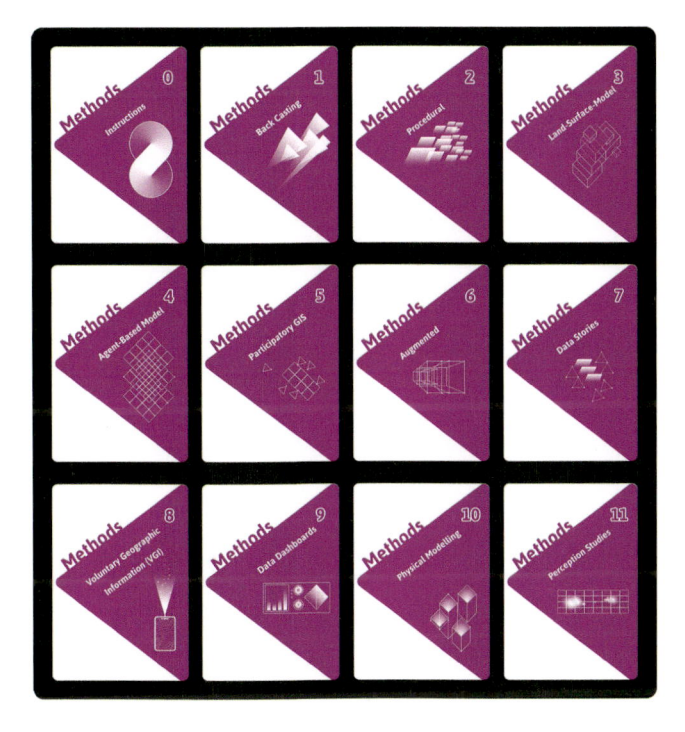

of people, models, and platforms (Figure 4.2). The purpose is twofold. First, it will continue with the framework on geodesign from Chapter 3, and second, it will offer a structured approach for readers to consider in their implementation of UDTs. In addition, some tables highlight various tools to support researchers in implementing these methods. The pace of change means that the methods described here are introductory rather than exhaustive and provide an overview rather than delving into specific technical details. A wide variety of data, platforms, and open-source software are available, several of which are listed here. Where possible, figures are used to illustrate practical examples of the mentioned methods, providing follow-up reading and considerations for future urban digital twin design. The rapidly evolving multiplicity of approaches to planning and geography, working with data, and modelling choices are also significantly influenced by new artificial intelligence (AI) techniques and the process of digital transformation and technocratic changes to planning

GEODESIGN, URBAN DIGITAL TWINS AND FUTURES

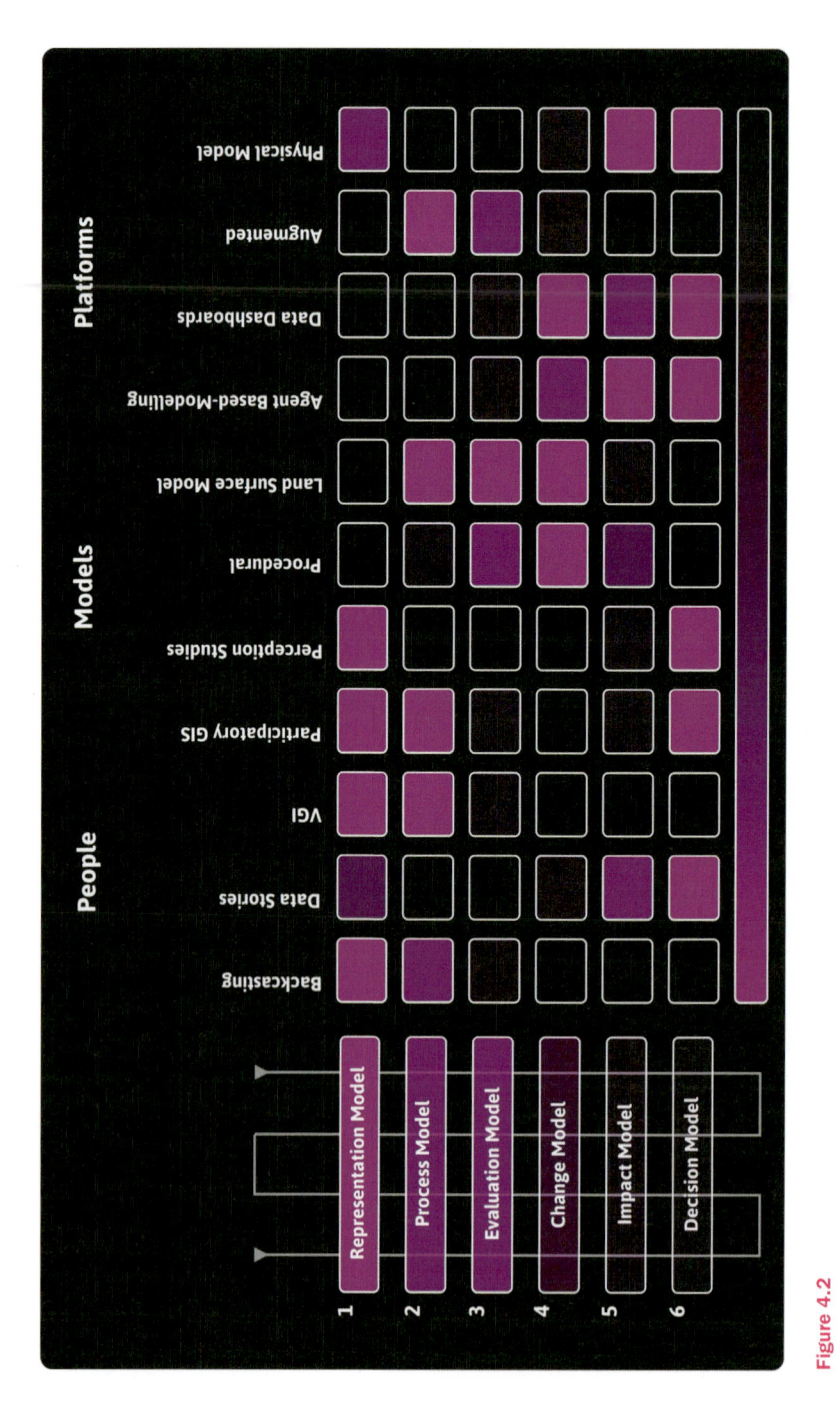

Figure 4.2
Authour, Geodesign Method Cards, and Suggested Matrix of Application to Geodesign Phases, 2024.-

systems (Birkin et al., 2021). Geodesign outcomes often focus on collaborative scenarios, so the methods selected should aim to provide reliable predictions of change when a specific design solution is used. It's important to note that in the first phase of geodesign, *The Representation Model*, data sources can be selected, and any software tools used in the project should be considered at this stage.

GEODESIGN METHODS – PEOPLE, MODELS, PLATFORMS

The following section describes 11 methods relevant to implementing urban digital twins (UDTs), particularly focusing on engagement, decision-making, and UDT data and analysis governance. These methods can be mapped to the six geodesign phases but are applicable across all steps (Figure 4.2). Readers should follow each

Figure 4.3
Authour, Geodesign Gameboard, 2023.

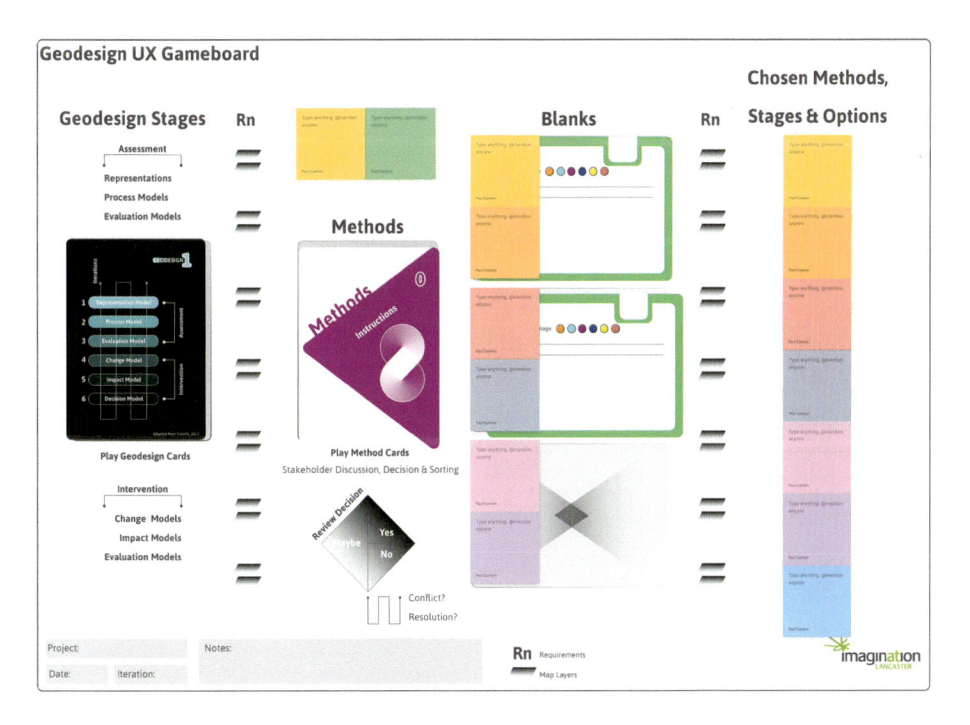

section's description and references, use the geodesign method board, and discuss the suitability and feasibility of their project. Ultimately, UDTs will heavily rely on a broader urban or city vision and a foresight process. This process allows the scope of the UDT and necessary transitions and utilises the urban visioning process and innovation (Dixon et al., 2023) (Figure 4.3). A geodesign card deck and game board can be used for specific research development and collaborative discussion.

PEOPLE

Backcasting – Geodesign Stage – Representation Model, Process Model

Futuring approaches in the built environment utilise scenario modelling to consider preferred futures and options. The selection of methods to achieve these preferable futures requires various iterations, and such a process can help projects progress towards the desired goals while proactively considering the urban context. These methods involve collaborative planning and dealing with uncertainty to establish a range of goal-based scenarios in which stakeholders work (Mäntysalo et al., 2023). Collaborations often use backcasting, which involves looking backwards in order to look forward to mapping desirable futures. Backcasting is valuable for generating collaborative insight into urban issues to develop future projects. The term "backcasting" emerged from Lovins (1977) and was developed by Robinson (1982) for energy backcasting in policy analysis.

Backcasting is an important method that allows for the addressing of sustainability issues in planning. Various futuring practices are often employed, and backcasting can be seen as a variation of Joseph Voros' Futures Cones (2003, 2017). This approach plots possible, plausible, probable, and preferable futures across set periods. The preferable futures are represented in the futures cone, which can be combined with a reverse cone that looks backward and forward. Simon Bibri has applied backcasting in an analysis of ecologically smart cities through four cases categorising cities: (1) Gothenburg and Helsingborg as compact cities, (2) Stockholm and Malmö as eco-cities, (3) London and Barcelona as data-driven smart cities, and (4) Stockholm and Barcelona as environmentally data-driven smart, sustainable cities (Bibri, 2020). Backcasting can be utilised across a range of urban indicators, but this categorisation approach, which Bibri uses, can aid future scenarios and considerations of sustainability goals. There are a number of variations to the backcasting approach, which should also be noted (Dreborg, 1996). These plots are established through historical analysis of patterns and events across a time period to identify the necessary change values to direct actions towards the plausible futures envisaged. Importantly, as Simon Bibri states,

> **The back casting approach takes into account the indeterminacy of the future and tries to define a broader conceptual framework for discussing the future; the study is less vulnerable to unforeseen change.**
> (Bibri, 2018, p. 12)

The required qualitative analysis may involve historical visual materials, statistical data projections, and computational studies and scenarios. Regardless of the media, what is important is that backcasting, as a methodological approach, addresses and actively shapes ICT aspects of UDTs.

In Figure 4.4, Andy Hines et al. create one such variation of backcasting termed 'Transition scenarios' when existing scenario approaches are already in place, in this case for 2090, but require adjustment via backcasting to 2035 as this was deemed a period in which existing

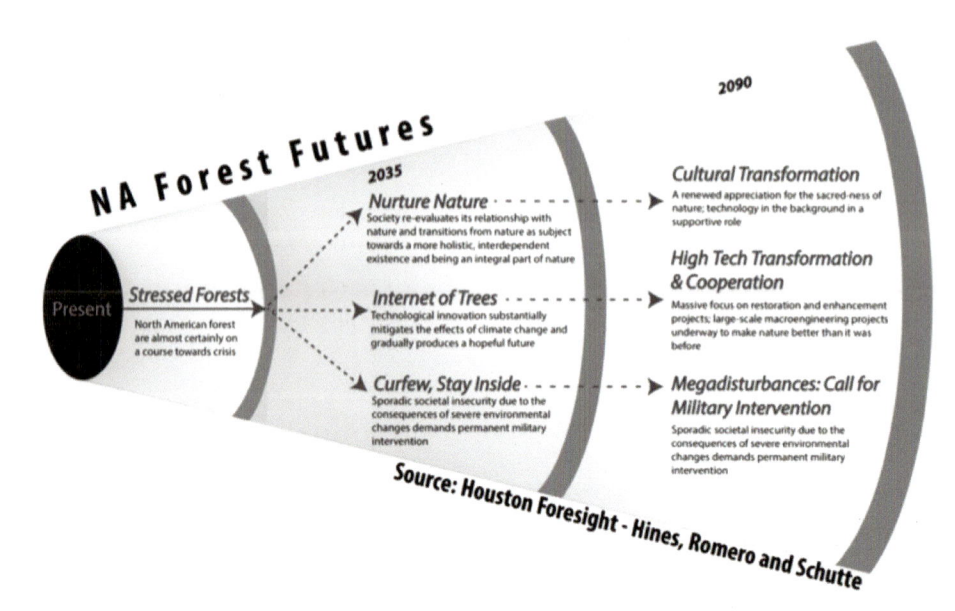

Figure 4.4
Any Hines, Johann Schutte, Maria Romero, North American Forest Futures Backcasting Scenarios. Journal of Futures Studies.

scenarios break down and alternative futures emerge. In this project for the University of Houston Foresight Program alongside the USDA Forest Service, the Strategic Foresight Group conducted horizon scans around the future of North American forestry for 2090. The diagram shows the methodological approach of offering a transition period in 2035 in which three scenarios are assembled, which map to the pre-established futures of 2090.

To implement backcasting, it is necessary to establish the time frame for historical research. Future objectives should be developed through a visioning process. Researchers must analyse the past to forecast the necessary steps towards that specific future, accounting for any technical differences and identifying the key changes and paradigm shifts. However, stakeholders' composition should be carefully evaluated for its representation and the dangers of narrow social dynamics that promote marginal preferences.

Data Stories – Geodesign Stage – Representation Model, Impact Model, Decision Model

Data stories utilise GIS data and maps to create compelling narratives about the study area, but they can also be connected to backcasting stages to narrate futures. The construction of data stories can involve archival imagery, video, and other media to tell the unique stories of a place. The design of the data story could use tangible and intangible data, such as intangible cultural heritage as defined by UNESCO (see table). The purpose is twofold. The first purpose of the method is to communicate complex phenomena in simple terms. The second purpose is to 'humanise' large-scale datasets and statistics through tangible narratives (Ash et al., 2024). Data stories are thus the visualisation and narrative of complex data between quantitative and qualitative approaches and can be created individually or nationally. However, creating data visualisation raises ethical questions regarding origins bias and feminist perspectives (D'Ignazio & Klein, 2020). Adam Frost, Tobias Sturt, Jim Kynvin, and Sergio Gallardo define four stages of storying data visualisation: *find, design, make, and refine*. The find stage is the story setting, design is setting the context and audience and setting the information hierarchy, and making involves design elements to create a consistent visual language. Refining involves testing and feedback (Frost et al., 2022, pp. 13–24).

The output of data stories could include creative mediums such as videography, ArcGIS StoryMaps, or other media. ArcGIS StoryMaps were created in 2019 as a cloud-based platform, free upon registration, which allows users to tell stories via maps and other media, such as the heritage action zone model and data story (Figure 4.5). The platform is important in consolidating a wide variety of media and bridging unstructured data and map representations, and its education and communication context has been assessed across a number of fields (Bartalesi et al., 2023).

Figure 4.5
Eleanor Brown & Paul Cureton, Lancaster, UK, Heritage Action Zone (HAZ), Terrestrial Laser Scan and Story Map for Augmented Experiences, Lancaster City Council, ArcGIS Story Maps, 2024. Supported by Digital Planning.GOV.UK.

Data Stories

Data Visualisation Europa Team	https://data.europa.eu/apps/data-visualisation-guide/
Andy Kirk, Visualising Data	https://visualisingdata.com/resources/
UNESCO, Intangible cultural heritage	https://ich.unesco.org/en/dive

VGI – Geodesign Stage – Representation Model, Process Model

The use of volunteered geographic information (VGI) from stakeholders and wider public donating street view imagery via smartphones (Chapter 1) or via engagement platforms such as Houseal Lavigne's Map.Social can be utilised to construct alternative datasets, create data stories, and be used as temporal data for backcasting. The use of VGI can mark out alternative objects and perceptions, validate other data sources, and monitor whether the frequency of updates is sufficient. Several platforms exist for collaborative mapping, including Open Street Map (OSM). VGI approaches are sought to address a digital divide between professionals and the public to analyse near real-time data, events, and spatial references, enabling decision-making and democratising data. For example, Alexander Dunkrel et al. (2023) explored social media reactions to georeferenced sunrise and sunset images via Flicktr and Instagram to analyse landscape preferences. In other cases, mobile phone sensors featured 95 participants to record mobility and urban barriers, visualised in a game environment (Helbig et al., 2022). The range of VGI work was part of a wider German Research Foundation Priority Program to explore VGI at various scales and across disciplines (2017–2022) (Burghardt et al., 2024).

Volunteered mapping and street view imagery can contribute towards modelling packages. Winston Yap (Yap et al., 2023) developed an urban network package to describe rich features across geographical scales. Utilising open street maps and Mapillary, a Python package was developed to map flows and connections for cities that can be overlaid and compared. Such lightweight tools allow a range of analytics, such as greenspace distribution and demographics, and advance network understanding for cities (Figure 4.6) (Yap & Biljecki, 2023). Junia Borges et al. have explored crowdsourced imagery for geodesign across a number of stages, but particularly around stakeholders, in the São Luiz and São José neighbourhoods in the Pampulha region in Belo Horizonte, Brazil. Voluntary information was gathered to map urban values by looking at security, noise,

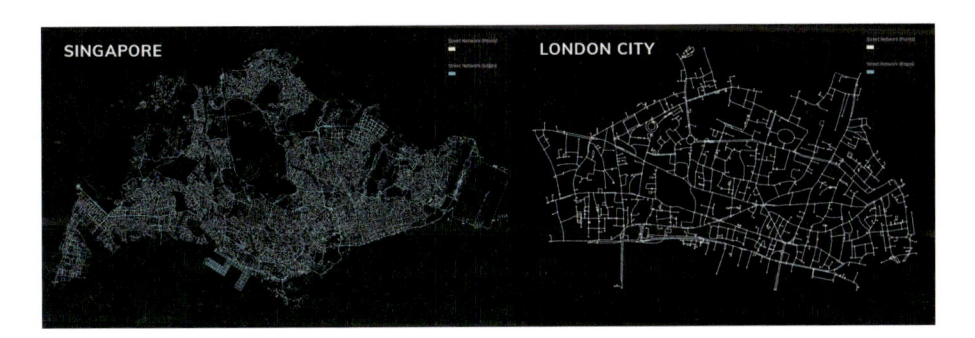

Figure 4.6
Winston Yap, Urban Analytics Lab, National University of Singapore, Urbanity – automated modelling and analysis of multidimensional networks in cities, 2023. https://github.com/winstonyym/urbanity.

and waste (Borges et al., 2015). The use of GPS smartphones, georeferenced points, objects, and attributes, labelling of maps, and or georeferenced imagery such as that found on Flickr provides an abundance of big data across a wide variety of activities (Goodchild, 2007). There are variations of VGI approaches, such as contributed geographic information (CGI) accounting for the source of generation and potential bias (Harvey, 2013). VGI use as a method raises issues around data quality, such as resolutions and coordinates, and ethical considerations, such as accidental submission of identifiable data, structure, and copyright, amongst many other considerations. However, VGI has the potential to address data inequality by reporting information-rich geographic features in low-mapped spaces and empowering communities to respond to environmental change through bottom-up processes (Sui et al., 2012). As geodesign seeks to facilitate dialogues across various actors, VGI has real potential to contribute datasets as part of a broader UDT and at the Representation and Process Model stage of the geodesign process.

Participatory GIS – Geodesign Stage – Representation Model, Process Model, Decision Model

Participatory GIS is an approach to gathering underrepresented groups and community knowledge on the study area using a GIS platform that can inform data stories and backcasting. Participatory GIS is often framed as a public participation geographic information system (PPGIS) and covers wide fields. VGI concentrates on the production, use, and sources of participant data, while PPGIS concentrates on participation in broadening planning perspectives and stages (Huck et al., 2014). Sarah Gottwald et al. have argued for the use of PPGIS methods in geodesign,

Combining PPGIS methods (instrumental approach) with geodesign (deliberative approach) holds great

> ***potential for spatial planning, as both approaches are based on georeferenced maps and assume the prevalent importance of place-based values.***
> (Gottwald et al., 2021, p. 1043)

Such an approach enriches the geodesign decision-making process. Participatory GIS can help develop trust and rigour in the impact and decision-modelling phases of geodesign. Participation as an instrumental approach can also include multidisciplinary teams. The method as an umbrella can involve analogue and digital mapping, interviews, VGI, and gaming (Chapter 5). Alexander Wilson and Mark-Tewdwr-Jones define three interleaved components in terms of public participation, separate from the public consultation (top-down) approach, engagement, and digital planning and have mapped at length histories and challenges of planning engagement as well as detailed approaches for consensus building and participation (2021). One project by Wilson and Tewdwr-Jones involved HCI in creating JigsAudio (Figure 4.7),

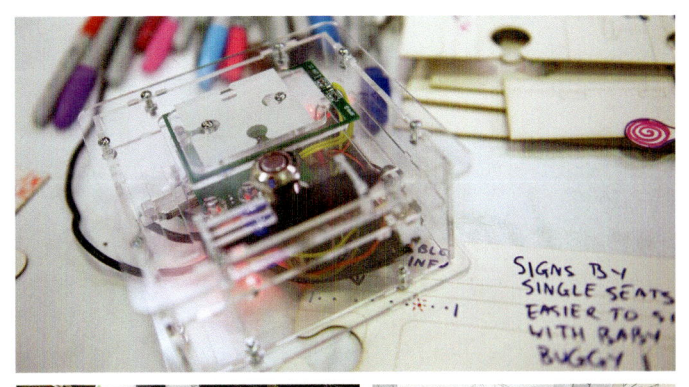

Figure 4.7
Alexander Wilson, Mark Tewdwr-Jones, Jigs, JigsAudio, Audio Device; left: Let's Talk Parks Deployment; right: Aliens Love Underpants Deployment. http://jigsaudio.com/.

an open-source replicable device that encourages drawing and recording audio simultaneously. Participants draw on a large card or puzzle piece and then place this on a radio-frequency device (RFID), allowing users to write, draw, and speak simultaneously. The device was used to record people's experiences and judgements of new metro carriage designs for the city of Newcastle, UK (Wilson and Tewdwr-Jones, 2020).

In the County of London Plan, published in 1943 and written by Patrick Abercrombie and John Forshaw, the future of London is presented, including the mapping of community clusters, defined as neighbourhood units. In Figure 4.8, existing community infrastructure is mapped. Abercrombie's plan for London was significant, both in setting the future growth for greater London, including satellite towns and garden cities, but also in the social make-up of the city. In Figure 4.9, the London Planning DataHub, Mayors Office, Greater London Authority, is a collaborative project bringing data from various indicators across all London authorities. Various classifications similar to the original London Plan of 1943 can be observed in the planning data map. Of the 44 million calls on the Greater London Authority held spatial data in 2023, 11 million were on the planning data map (up from 6 million in 2022). In this case, using dashboards (via Kibana) and web-based cloud mapping has massively increased the engagement aspect of planning mechanisms for London.

Applying participatory GIS should ultimately consider what Eveliina Harsia and Pilvi Nummi term 'polyphony', that is, incorporating multiple voices and co-presence within a planning process. However, this can be compounded by digital skills divides and the co-design process. In an applied case, this was demonstrated in Kontula, east Helsinki, engaging migrants through a PPGIS questionnaire and visualising responses (Harsia & Nummi, 2024).

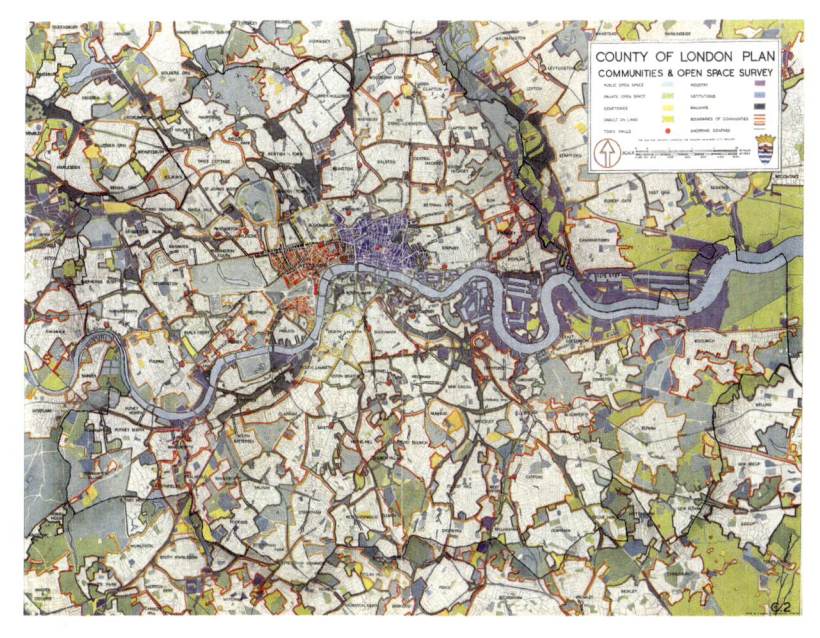

Figure 4.8
The Communities and Open Space Survey, published in the London County Council's County of London Plan, drawn by Arthur Ling, D.K. Johnson, and G. Hannah in 1943. Alamy Stock Photo.

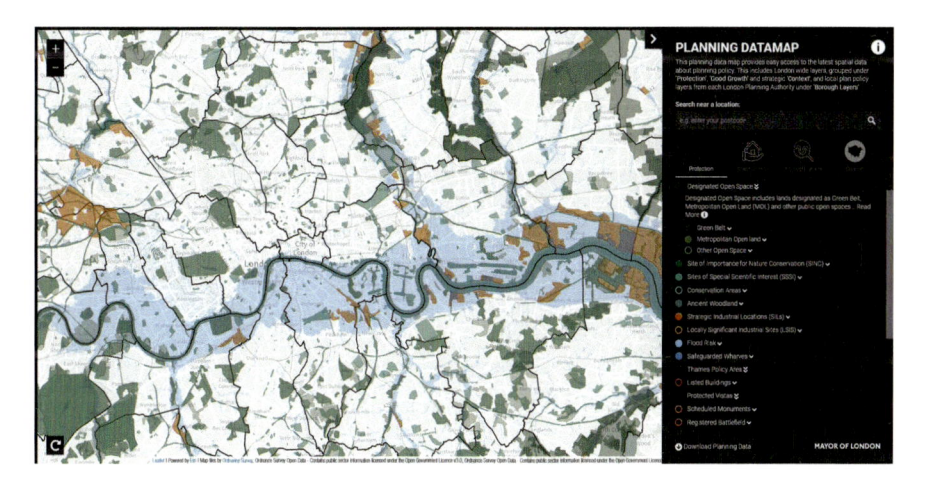

Figure 4.9
Greater London Authority, Planning DataMap, 2023. Esri and map tiles by Ordnance Survey. This planning data map provides users with the latest spatial data about planning policy. Three headings are used. 'Protection' and 'Good Growth' show planning designations and how development is regulated. 'Context' shows strategic policies for development in London. https://apps.london.gov.uk/planning/.

MODELS

Perception Studies – Geodesign Stage – Representation Model, Decision Model

Perception studies are perceptions and values of the urban environment that are used as models to aid decision-makers for regeneration and transformation and can be connected to *people*: backcasting, data stories, and VGI. Perception models can be designed via various options, such as eye-tracking, A/B tests, pairwise comparisons, or semi-structured interviews. Perception studies typically measure visual material such as street view images or maps or, in digital settings, virtual replicas of environments via eye tracking or user interface prompts measuring things such as urban walkability.

Kevin Lynch's Image of the City (1960) sought the city's imageability through extensive interviews, photography, and map sketches in order to provide five underlying structures of a place, which he termed 'Paths, Nodes, Edges, Districts and Landmarks' and still remains a primary method and approach for decoding city perception (Figure 4.10). Lynch was one of many other urban design practitioners seeking to understand the city's legibility, including Jane Jacobs, Christoper Alexander, and Gordon Cullen. Perception studies seek to understand the legibility of the environment through aesthetic, observational criteria in situ or through evaluation of visual and GIS-based material. A/B tests involve user experience evaluation of two or more random variants often in design research measuring aesthetic preference; pairwise comparisons compare pairs head-to-head and can be evaluated using computational and qualitative techniques in the field or in lab settings often used as part of analytic hierarchy processes (AHP) to measure hierarchy. In one case, AHP and pairwise comparisons were used to evaluate notions of sustainable neighbourhoods (Guillén-Mena et al., 2023). Stakeholders can explore visual preferences of design options and generate heat maps of eye patterns and visual focus using eye-tracking devices.

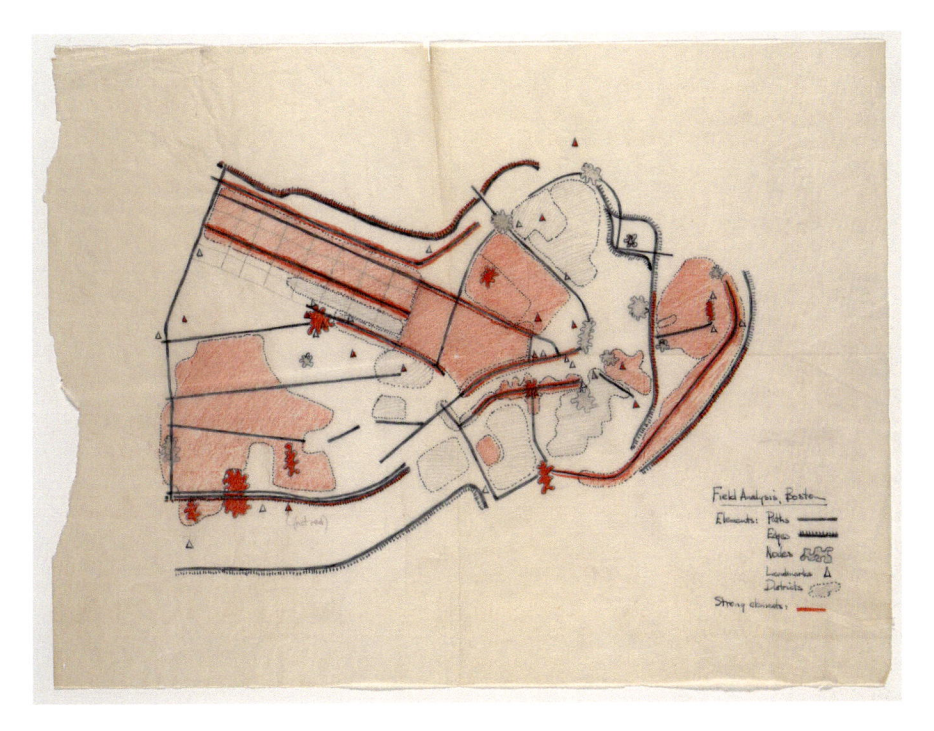

Figure 4.10
Kevin Lynch, Extract from *The Perceptual Form of the* City, Boston, Massachusetts, 1954–1959. Image courtesy of the MIT Archives. © Copyright 2014. Lynch's hand-drawn map results from an extensive survey of participants' perceptions of the city, interviews, photographs, and field trips. Lynch famously classifies nodes, paths, edges, districts, and landmarks as shared and common perceptual city forms (*The Image of the City*, 1960).

Pairwise comparisons are a lab-based approach to evaluating environments, often assessed via eye-tracking or survey prompts. A number of images are created as a sample, and users compare through an iterative cycle based on set evaluation criteria. A computer vision algorithm by Abhimanyu Dubey et al. called Place Pulse (1.0), created a crowdsourced game prompting users to evaluate street view imagery in response to questions 'Which place looks safer,' 'Which place looks more unique,' and 'Which place looks more upper class?' to which 200,000 comparisons resulted from 4,109 images of New York, Linz, and Salzburg, though dependent on participant demographics for representation. However, this was not scalable to global cities and resulted in a new dataset, Place Pulse (2.0),

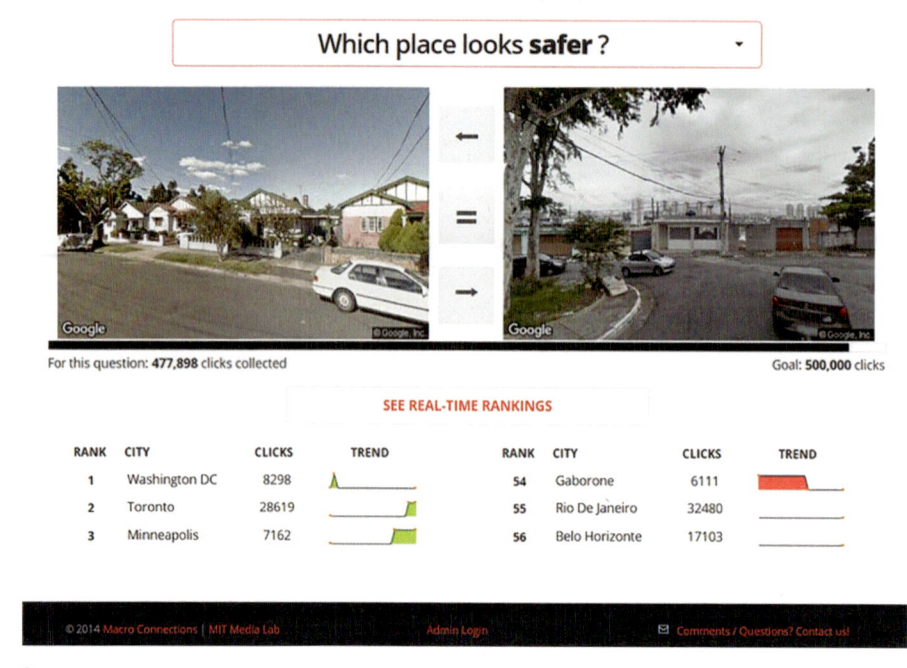

Figure 4.11
Abhimanyu Dubey, Nikhil Naik, Devi Parikh, Ramesh Raskar and César A. Hidalgo, 2016. Example results from Place Pulse 2.0 ranked images based on pairwise comparisons.

in which 1.17 million pairwise comparisons for 110,998 images, 56 countries from 81,630 online visitors evaluated street view images across the parameters of 'safe, lively, boring, wealthy, depressing and beautiful' (Dubey et al., 2016, pp. 197–198) (Figure 4.11).

Pairwise Comparisons

Place Pulse 2.0   https://paperswithcode.com/dataset/place-pulse-2-0

Procedural – Geodesign Stage – Evaluation Model, Change, Model, Impact Model

In order to generate the preferable futures of stakeholders via back-casting, narrate future conditions via data stories, or transform the built environment via VGI, procedural modelling approaches offer future-based scenarios at speed. They can be combined with many of the methods discussed here. ESRI CityEngine is an example of a modelling L-system (used to model the morphology of things) that ESRI R&D Center, Zurich, originally developed. CityEngine was formed by Yoav Parish and Pascal Müller over 20 years ago (Müller et al. 2006; Parish & Müller, 2001). CityEngine is particularly important for modelling road networks and (parametric) buildings, both contemporary and historical (Saldaña, 2015; Hosseini et al., 2022). CityEngine geometries can be controlled via programmed rules and or rulesets called computer generated architecture (CGA) shape grammar. The grammar and rulesets allow control of elements such as floor heights or building orientation via a user interface. In addition to this capability, CityEngine can incorporate a variety of GIS datasets for real-world mapping (Figure 4.12). Procedural modelling, as a method, provides a benefit through the iterative approach possible via variations of the CGA at speed. The rule and ruleset-based approach allows the generation of a variety of possible 3D/4D

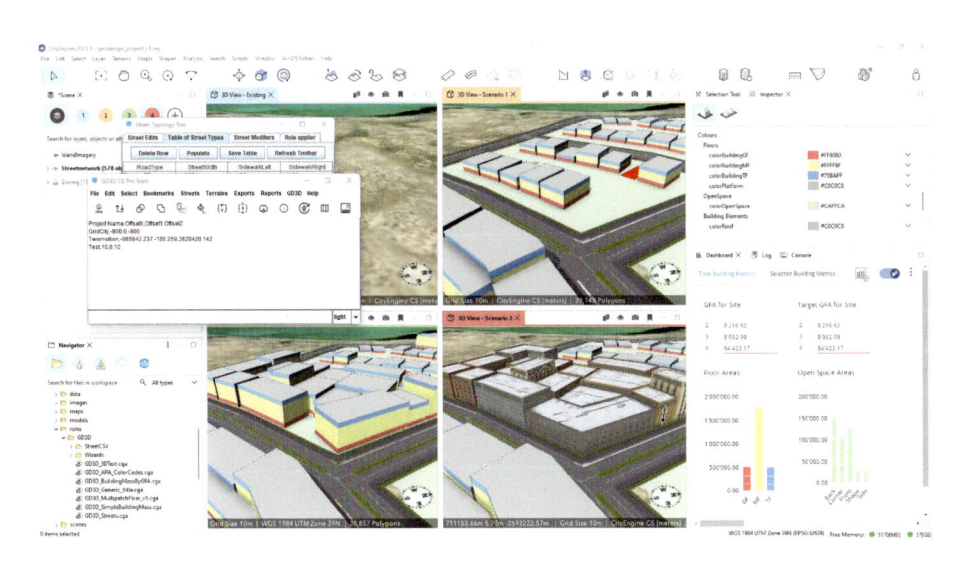

Figure 4.12
Elliot Hartley, CityEngine Procedural Modelling, 2024.

scenarios to be generated and virtualised. In addition, rules-based approaches allow specific constraints to be set to any potential buildings and urban blocks, mimicking planning conditions such as zoning or amenities values. Procedural modelling allows the setting of parents and control of any sub-variants to a particular building or block for finite adjustment. There are stark variations compared to other modelling choices, such as non-uniform rational B-spline (NURBs) or polygonal modelling, which can require extensive manual editing and modelling. These modelling programs include Blender, MeshLab, Houdini and Rhino, and Grasshopper. However, this form of production is changing with procedural city plugins for Blender, AI via prompts, and image-to-model algorithms, which will increase the accessibility of these software programs and the capability of planning professionals with clear planning policies. Overall, procedural modellers, such as CityEngine, offer a range of benefits to urban planning but also the movie and games industry production pipelines and have been part of the production workflow in series such as The Witcher, Netflix (2020) and films such as BladeRunner 2049, Warner Bros (2017) (Figure 4.13).

Figure 4.13
BladeRunner 2049. Directed by Denis Villeneuve and produced by Ridley Scott, written by Hampton Fancher and Michael Green. Based on Do Androids Dream of Electric Sheep? by Philip K. Dick. Warner Bros. Alamy Stock.

Flora Roumpani's work matched CityEngine to the QUANT model (Chapter 1) to assess housing capacity by providing population and employment matrix inputs. Subsequent supply and demand then inform the CGA rules and allow a planner to make decisions regarding infrastructure (Roumpani, 2013, 2022, p. 325). Roumpani's previous work in the project ReMap Lima involved mapping via drone acquisitions and procedural modelling via CityEngine, which was used as an important demonstration of urban expansion via procedural approaches (ReMap, 2014). This procedural method should be selected to create 3D urban models of various development options and automate planning proposals based on real-world planning restrictions but does not model social systems. Procedural modelling could also automate large-scale urban layouts and structures with various constraints and optimisations, generate rapid prototypes and visualise results at speed, and provide large context models that transfer to game engines, allowing XR interactions and public participation.

Land Surface Model – Geodesign Stage – Process Model, Evaluation Model, Change Model

Land surface models (LSMs) represent soil, vegetation, energy, water, and atmosphere exchanges. They concentrate on climate or earth systems and are used to understand the dynamics of land surface and atmosphere under climate change. This modelling method incorporates various processes (components) in quantitative forms, such as water bodies or vegetation dynamics. LSMs represent and model the complexity and dynamics of the Earth's surface for future trajectories and land heterogeneity. LSMs can measure exchanges such as rainfall runoff and carbon from leaf litter transferred to soils, which are essentially bio-geo-chemical exchanges (Figure 4.14). An example of an LSM is the Joint UK Land Simulator (JULES), which is a unified model of global climate change to local climate as part of a unified earth system. JULES and the unified model are part of the

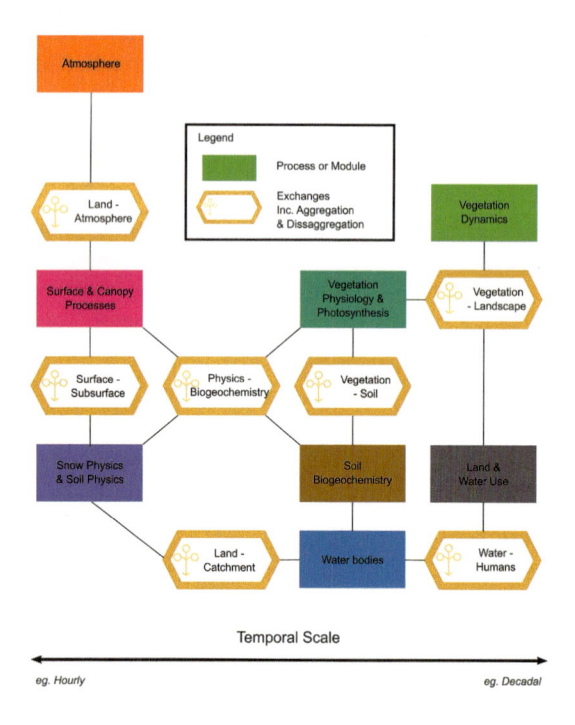

Figure 4.14
"Schematic of Land Surface Model showing Components (process or module) and the exchanges between components across temporal scales (hourly to decadal)" produced with permission from (Blyth et al., 2021, p. 47).

UK's MET office climate and weather prediction strategy. However, some LSMs do not have urban schemes and components in a unified climate model, making simulations of cities more problematic regarding detailed site information and surface types (Lipson et al., 2024).

LSMs require modellers to communicate risk and set temporal timescales to stakeholders in the geodesign process, to which land surface processes can be communicated and robust change models created. Jianguo Wu proposes a sustainability approach combining aspects of environmental science for operationalising land use change and geodesign (Huang et al., 2019; Wu, 2019). LSMs are increasing in capability for simulation of various complex exchanges and have high methodological importance, though the methodological challenge is the science communication of land surface model outputs describing the atmospheric and physical changes to land from climate models and combining LSMs in the geodesign process.

Land Surface Models	
JULES	https://jules.jchmr.org/
Community Land Model	https://www.cesm.ucar.edu/models/clm
NOAH-MP	https://ral.ucar.edu/model/noah-multiparameterization-land-surface-model-noah-mp-lsm

Agent-Based-Modelling – Geodesign Stage – Change Model, Impact Model, Decision Model

ABM is a computational modelling process that understands agents and systems (e.g., people) and their behaviours for advanced simulation of future actions under different environmental parameters. ABM can be used alongside procedural modelling and the generation of 'what-if scenarios'. ABM approaches allow the mapping of emergent behaviours between agents, allowing a range of modifications and simulations (Batty, 2007). Several platforms exist, including Gamma and Netlogo (Taillandier et al., 2019). Various scenarios can be played out, which leads this modelling approach to geodesign in terms of the approach's forecasting ability. Combined with environmental data, GIS and embedded sensors linked to ABM models, individual actors, and large population behaviours can be mapped for interactions (Amos & Webster, 2022; Amos et al., 2023). As a modelling method, ABM relies on focused research questions and is rarely coupled to other models of various complex aspects of urban spaces. ABMs also require careful validation and evaluation (Crooks et al., 2019; Crooks et al., 2021).

The City Scope project by MIT Media Lab contains several open digital platforms for a variety of urban issues (Figure 4.15). A module-based framework allows standalone analysis of different configurations. An additional tool also maps interactions on a Lego grid networked to projects. Future potentials have been defined by Nick Malleson et al., who argue that we could see wider use of ABM for large-scale pedestrian simulation in near real time, what is called

Figure 4.15
City Science, City Scope, 2016–, MIT Media Lab. https://cityscope.media.mit.edu/.

full-probabilistic, which accounts for uncertainty in mobility, such as individual route choices by cyclists or drivers, which requires multiple emulations of possible choices, which could be used for policy and governance via a digital twin approach. However, as the authors assert, there are still remaining challenges around ABM scales, complexity, rigour, and ethics of the data used in ABMs (Malleson et al., 2022). ABMs would be appropriate for change and impact models to create what if scenarios in geodesign for UDTs.

Agent-Based Modelling

Gamma Platform	https://gama-platform.org/wiki/Home
Repast	https://sourceforge.net/projects/repast/
Netlogo	https://ccl.northwestern.edu/netlogo/
MASON	https://cs.gmu.edu/~eclab/projects/mason/

PLATFORMS

Data Dashboards – Geodesign Stage – Change Model, Impact Model, Decision Model

Data dashboards are visualisations of complex environmental data and metrics to inform decision-making processes. Urban dashboards stem from early mission room control services and decision support tools (Mattern, 2015). Early prototypes include CASA's

CHAPTER 4
GEODESIGN METHODS FOR URBAN DIGITAL TWINS

City Dashboard in 2012 (Gray et al., 2016). Careful dashboard design is required and has ethical considerations. Dashboards are presented in terms of simple, readable graphical representations of a range of static or dynamic urban indicators, such as the Smart City Amsterdam dashboard by Geodan (Figure 4.16). Dashboards

Figure 4.16
Geodan, Sogelink Group, Amsterdam Dashboard, 2020. Geodan produced an open Smart City Amsterdam dashboard to analyse energy transition, living environments, and mobility for 2050. The dashboard could also be integrated into BIM and GIS platforms. It shows real-time traffic flows and energy consumption and distribution.

are a primary medium for distributing and communicating research findings or governmental data to other agencies and organisations and the public, such as the London Data Hub. Dashboards are a critical component of UDTs but are also a way of presenting stages in the geodesign process, especially for change and impact models. As Oliver Dawkins et al. account, reducing a variety of spatial and numerical indicators into one screen, or often a 'control room' management suite, raises questions about validity and trustworthiness. These issues were apparent in implementing the Dublin City Dashboard and using 3D modelling; the display is a synthetic landscape for the display of data. (Dawkins et al., 2021; Young & Kitchin, 2020). Shannon Mattern has also levied criticism on a range of dashboards for lacking experiential aspects around the origin of data and the aspects of immersing the viewer in the data in order to understand it (Mattern, 2017). Ultimately, how dashboards are framed is a key deliberation in geodesign and in how the UDT functions, is adopted, and informs decision-making. The dashboard is a fundamental data visualisation intertwined with user experience design for governance and service design aspects that need careful construction (De Lange, 2018).

In an analysis of dashboards across Germany by Christoph Huber et al. (2023), the authors sampled 42 cities and 16 public dashboards, and the most common graphical features are KPIs, column timelines, doughnuts, indicator matrices, and dot density graphs. The graphical features themselves are critically important in terms of viewers' reception and 'reading' ability. Such a focus is necessary for the work of Maria-Lluisa Marsal-Llacuna, who has suggested a people's smart city dashboard which seeks to establish collaborative citizen-centric metrics (Marsal-Llacuna, 2020). Data dashboards must provide legible and transparent metrics, be accountable, and not just enforce one-way data flow but provide civic empowerment and support collaborative decisions in the geodesign phases.

CHAPTER 4
GEODESIGN METHODS FOR URBAN DIGITAL TWINS

Dashboards

London, UK Data Dashboard	https://london.datahub.io/
Chicago, USA	https://www.chicago.gov/city/en/sites/community-safety/home/data-dashboards.html
Hong Kong	https://data.gov.hk/en/city-dashboard

Augmented – Geodesign Stage – Process Model, Evaluation Model

Extended reality (XR) incorporates virtual, mixed, and augmented reality devices that have been applied to urban simulation, analytics, and visualisation (Augmented Urbans, 2019). XR has been applied across community planning, urban planning, smart city research, human-computer interaction, architectural research, spatial cognition, and wayfinding, and it has a direct relationship with the games industry (Cureton, 2022). Milgram and Kishino's display taxonomy is a strong base for understanding the continuum of XR, which presents a condition for virtual and mixed display based on knowledge about the world, the fidelity of reproduction, and the extent of user presence (1994). The synergy between 3D GIS and porting to game engines for XR, such as UNITY or Unreal Engine, allows the development of various interactions and triggers for built-environment experiences played in first-person (FPV) mode.

However, augmented methods do not just involve immersion within an environment but augment urban dynamics. In the Urban Toolkit (Figure 4.17), 'what if' scenarios can be created through a web-based 3D visualisation tool through a concept called *Knots*, which merges thematic data with physical data by creating grammar-based ties between linking layers, for example, sun shading maps draped over buildings (Moreira et al., 2023, p. 2). Users input several layers, including open street maps or CSVs, and the tool creates an accessible platform for 3D analytics compared to premium commercial software and often the extensive training required. This web-based

199

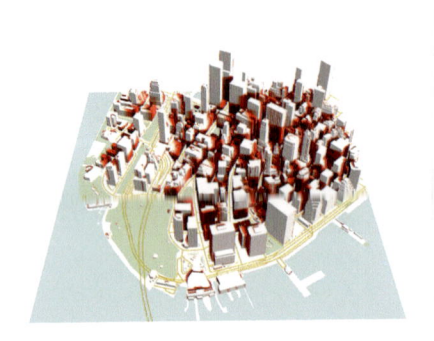

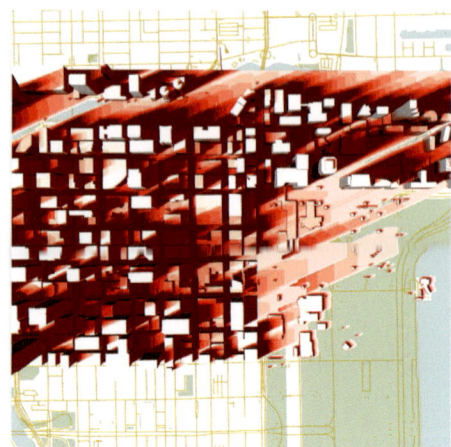

Figure 4.17
Gustavo Moreira, Maryam Hosseini, Md Nafiul Alam Nipu, Marcos Lage, Nivan Ferreira, and Fabio Miranda, UrbanToolkit, 2023. https://urbantk.org/.

tool augments various scenarios and impacts and is useful for the process and evaluation model stages of geodesign.

While the digital twin of Herrenberg, Stuttgart, Germany, did not use the geodesign process, its participatory nature is pertinent to support the hypothesis that the geodesign framework lends itself to emergent UDTs. In the Herrenberg digital twin, researchers at the High-Performance Computing Center Stuttgart embedded IoT, 3D modelling, and simulation and utilised virtual reality for widespread public and professional engagement. The Herrenberg twin contained various assembled models, including a space syntax road network. From this basis, the researchers utilised CAVEs (cave automatic virtual environment) and mobile VR set-ups to engage citizens and validate models using empirical data and interviews with a small participant sample (Dembski et al., 2020, pp. 10–13). Through establishing the twin of Herrenberg, a number of related research projects could be undertaken. A range of stakeholder groups was consulted in different planning schemes, such as new retail offerings and transport planning (Dembski et al., 2020). High Performance Computing Center

Stuttgart (HLRS) has continued using the twin as a critical platform in the CapeReviso project to develop planning and support tools for data-driven decisions around cyclists and pedestrian conflict, urban design, and traffic planning. Augmenting various scenarios as a consultation tool with stakeholders, including planners and the public, can aid in communicating geodesign change models, impacts, and decisions. This could involve virtual reality or augmented reality via CAVEs or mobile devices, which can be used to visualise various simulations and models. The Herrenberg twin is an excellent example of experimentation and engagement (Figure 4.18).

Andy Hudson-Smith has accounted for one element of the trajectory of XR in terms of a broader discussion of the metaverse (virtual world based on the novel *Snow Crash* by Neal Stephenson) promoted by Meta (Facebook), mirrored reality estates, citing early precedents

Figure 4.18
Dembski, F., Wössner, U., Letzgus, M., Ruddat, M., & Yamu, C., High-Performance Computing Center Stuttgart (HLRS), 2020. Herrenberg, Germany. This mobile VR demonstrates a participatory process that includes a back-projection wall, tracker, and 3D projector. Users interface via active shutter glasses. Photography, Dembski, 2019.

Figure 4.19
Second Life, Kawaii City, Linden Labs, 2023. A cyberpunk dystopian city world for players to explore.

such as the game Second Life by Linden Labs first established in 2003 (Figure 4.19) in which players meet in a multimedia platform in various collaborative worlds, which point to emerging projects seeking to establish "a collaborative, occupied virtual space where users can build anything, own land, edit, and inhabit the environment. Arguably, it is the future of digital urban planning; the hard part is building it" (Hudson-Smith, 2022, p. 348). The degree of immersion can vary, as hardware costs and user suitability in terms of navigation and usability from a range of demographics can impact it. The use of smartphones for augmented reality has also been explored in several digital twin cases, such as the 'Augmented Urbans – Visionary, Participatory Planning and Integrated Management for Resilient Cities project,' covering Latvia, Sweden, Finland, and Estonia. A number of XR cases and prototypes were explored, including AR AVALINN (Figure 4.20), to create a pollinator highway for biodiversity and the creation of a linear park in Tallin, Estonia. Augmented approaches can be used as part of an evaluative process or through interaction with process models.

CHAPTER 4
GEODESIGN METHODS FOR URBAN DIGITAL TWINS

Figure 4.20
Spatial Design Competence Center of the Tallinn Strategy Unit. AR AvaLinn, 2019. Eerik Kändler, Johanna Jõekalda, Meem, Multistab, Silver Seeblum. Photography by Johanna Jõekalda, Mats Õun.

GIS Plugins for Games Engines	
ArcGIS Maps SDK for Unreal Engine	https://developers.arcgis.com/unreal-engine/
Cesium for Unreal Engine / Omniverse / Unity	https://cesium.com/platform/cesium-for-unreal/

GIS & Urban Based Tools	
UrbanToolkit	https://urbantk.org/getting-started/
Global Urban Network Dataset	https://figshare.com/articles/dataset/Global_Urban_Network_Dataset/22124219 & https://urbanity.readthedocs.io/en/latest/
Gephi	https://gephi.org/
CapeReviso	https://capereviso.hlrs.de/

203

Physical Model – Geodesign Stage – Representation Model, Impact Model, Decision Model

The creation of physical models of options and scenarios helps quantify changes in the geodesign process. Models could be cardo, out foam, or plywood to create immediate iterations of the design ideas and the modelling of the environment (Moffitt, 2023). The use of 3D printing or other approaches, such as CNC routing, is particularly useful in large urban areas and existing settlements. The use of 3D printing via colour jet printing, selective laser sintering, and stereolithography allows a range of finishes, production timescales, and sizes. Printing choices affect the maximum print volume possible in any single print and the finish, which can be rough or smooth. Spray coating is also optional. Often, physical models are sprayed white in order that GIS-based data can be projected onto the model. Time-series data can be used effectively. Optionally, the model may be assembled as a kit of parts, and various parts can be replaced for updating, reflecting real-world planning development. Modelling choices should be selected based on the iterative stages of the geodesign process, and more detailed models should be selected for context needs and decision models. As a method, models fundamentally display the volume of the study area, often lost in digital environments and 2D representations.

The City of Munich, Germany, augmented its 3D city model for meetings with the Department of Urban Planning and Building Regulations through the use of hololens, removing the need for a physical model and increasing interaction with digital data in planning policy development. Virtual Singapore, as part of the Urban Redevelopment Authority (URA), has an extensive public exhibition space containing several interactive exhibits, including a central area model and an island-wide model, which is used as part of future master plans and

CHAPTER 4
GEODESIGN METHODS FOR URBAN DIGITAL TWINS

Figure 4.21
Pipers Model Makers, New London Architecture (NLA) Model, Photography Agnese Sanvito and Paul Raftery, 2015. The model highlights any design scheme that has recently been completed, is currently under construction or has planning permission. Current London planning projects are highlighted through 3D-printing technology and a projection system linked to an interactive digital display that contains full project details.

public consultation. Finally, Figure 4.21, the New London Model, has been established for over 40 years and has undergone various refurbishments at a 1:2,000 scale covering 195 km². In these cases, models are chosen as they embody the organisation's or institution's values and culture and are fundamental learning tools (Bekkering et al., 2020).

205

CASE STUDY – DATA CITY DUBLIN

Data City Dublin is a 3D printed model spanning 28 km² of the Irish capital at a scale of 1:2,000. Created by the Data Stories project at Maynooth University, its purpose is to help the presentation and

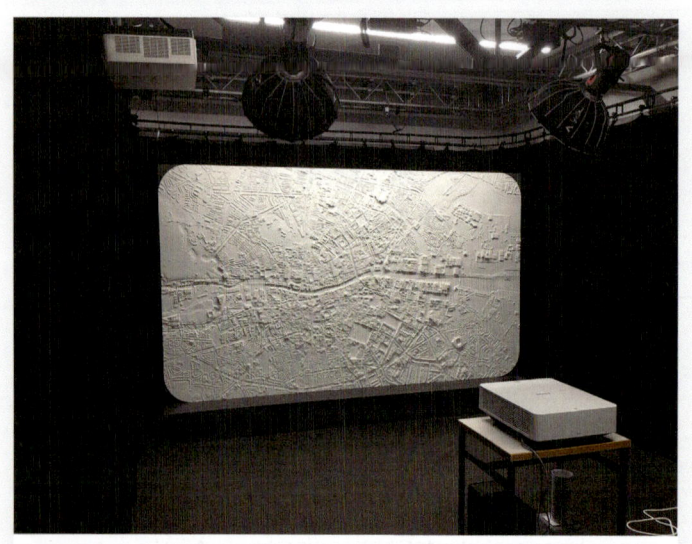

Figure 4.22
Olly Dawkins, Data City Dublin, 2023. https://datastories.maynoothuniversity.ie/.

assessment of evidence related to Dublin's current housing crisis in public exhibitions and forums. The 3D model provides a spatial reference which invites discussion as visitors share stories while they search to locate points of interest, such as their homes and places of work. When overlaid with data, the model serves to link personal narratives about place to the more abstract measures and classifications of official housing, planning, and property datasets. Time-series aerial imagery illustrates the city's evolution, while demographic, building use, and property price data contribute to a more multi-faceted overview of the situation. In this way, the model presents opportunities for dialogue that contribute to a more nuanced understanding of personal struggles over housing in relation to longer-term patterns and trends (Figures 4.22).

https://datastories.maynoothuniversity.ie/

SUMMARY

The development of UDTs can engage with a wide range of methods. There are no set paths to follow; rather, there is a series of flexible choices related to organisation, collaboration, modelling, participation, and consideration of future scenarios for the area in question. Geodesign is designed to be an adaptable, procedural, and iterative collaborative decision-making process that can be applied at different scales and in various locations. Therefore, the design of digital twins and their data approaches and models via a systems approach should also be highly adaptable and individual but also connected intrinsically to geodesign. Caldarelli et al. (2023) advocate for the use of complexity science in digital twins and propose a shift in perspective to see digital twins as living systems. This shift would recognise the need to consider living systems and self-organisation, requiring flexible methods in representation and modelling. The 11 methods mentioned here may not be directly related to each other, geodesign, or the current state of UDTs. However, they address broader questions about using UDTs in planning, focusing on three main areas: people, models, and platforms.

REFERENCES

Amos, M., & Webster, J. (2022). Crowdsourced identification of characteristics of collective human motion. *Artif Life*, 28(4), 401–422. https://doi.org/10.1162/artl_a_00381.

Amos, M., Gwynne, S., & Templeton, A. (2023). A dynamic state-based model of crowds. *ArXiv*. https://doi.org/10.1016/j.ssci.2024.106522.

Ash, J., Kitchin, R., & Leszcynski, A. (2024). *Researching digital life, orientations, methods and practice*. Sage Publications.

Augmented Urbans. (2019). Augmented urbans – Extending the urban planning practices. Accessed 20 Nov 2023. https://augmentedurbans.metropolia.fi/.

Bartalesi, V., Coro, G., Lenzi, E., Pagano, P., & Pratelli, N. (2023). From unstructured texts to semantic story maps. *International Journal of*

Digital Earth, 16(1), 234–250. https://doi.org/10.1080/17538947.2 023.2168774.

Batty, M. (2007). *Cities and complexity: Understanding cities with cellular automata, agent-based models, and fractals*. The MIT Press.

Bekkering, J.D., Curulli, G.I., & van Hoof, J.J.P.M. (Eds.) (2020). *Architectural models as learning tools*. Caleido.

Bibri, S.E. (2018). Backcasting in futures studies: A synthesised scholarly and planning approach to strategic smart sustainable city development. *European Journal of Futures Research*, 6, 13. https://doi.org/10.1186/s40309-018-0142-z.

Bibri, S.E. (2020). A methodological framework for futures studies: Integrating normative backcasting approaches and descriptive case study design for strategic data-driven smart sustainable city planning. *Energy Informatics*, 3, 31. https://doi.org/10.1186/s42162-020-00133-5.

Birkin, M., Clarke, G., Corcoran, J., & Stimson, R. (Eds.) (2021). *Big data applications in geography and planning: An essential companion*. Edward Elgar Publishing. ISBN 978-1-78990-978-4.

Blyth, E.M., Arora, V.K., & Clark, D.B. et al. (2021). Advances in land surface modelling. *Curr Current Climate Change Report*, 7, 45–71. https://doi.org/10.1007/s40641-021-00171-5.

Borges, J., Jankowski, P., & Davis, C.A. (2015). Crowdsourcing for geodesign: Opportunities and challenges for stakeholder input in urban planning. In C. Robbi Sluter, C. Madureira Cruz, & P. Leal de Menezes (Eds.), *Cartography - Maps connecting the world*. Lecture notes in geoinformation and cartography. Springer. https://doi.org/10.1007/978-3-319-17738-0_25.

Burghardt, D., Demidova, E., & Keim, D.A. (Eds.) (2024). *Volunteered geographic information interpretation, visualization and social context*. Springer Professional.

Caldarelli, G., Arcaute, E., & Barthelemy, M. et al. (2023). The role of complexity for digital twins of cities. *Nature Computational Science*, 3, 374–381. https://doi.org/10.1038/s43588-023-00431-4.

Council, G., & Lamb, K. (2022). *Gemini papers: What are connected digital twins*. Centre for Digital Built Britain. https://doi.org/10.17863/CAM.82194.

Crooks, A.T., Malleson, N., Manley, E., & Heppenstall, A.J. (2019). *Agent-based modelling and geographical information systems: A practical primer*. Sage.

Crooks, A., Heppenstall, A., Malleson, N., & Manley, E. (2021). Agent-based modeling and the city: A gallery of applications. In W. Shi, M.F. Goodchild, M. Batty, M.P. Kwan, & A. Zhang (Eds.), *Urban informatics. The urban book series*. Springer. https://doi.org/10.1007/978-981-15-8 983-6_46.

Cureton, P. (2022). Augmented reality: Robotics, urbanism and the digital turn. In *The Palgrave encyclopedia of urban and regional futures*. Palgrave Macmillan. https://doi.org/10.1007/978-3-030-51812-7_2 31-1.

Dawkins, O., Kitchin, R., & Young, G., (2021). City dashboards and 3D geospatial technologies for urban planning and management. In A. Rae, & C. Wong (Eds.), *Applied data analysis for urban planning and management* (pp. 83–102). Sage Publications.

De Lange, M. (2018). From real-time city to asynchronicity: Exploring the real-time smart city dashboard. In S. Lammes, C. Perkins, A. Gekker, S. Hind, C. Wilmott, & D. Evans (Eds.), *Time for mapping* (pp. 238–255). Manchester University Press.

Dembski, F., Wössner, U., Letzgus, M., Ruddat, M., & Yamu, C. (2020). Urban digital twins for smart cities and citizens: The case study of Herrenberg, Germany. *Sustainability*, 12(6), 2307. https://doi.org/10.3390/su12062307.

D'Ignazio, C., & Klein, L.F. (2020). *Data feminism*. MIT Press.

Dixon, T.J., Karuri-Sebina, G., Ravetz, J., & Tewdwr-Jones, M. (2023). Re-imagining the future: City-region foresight and visioning in an era of fragmented governance. *Regional Studies*, 57(4), 609–616. https://doi.org/10.1080/00343404.2022.2076825.

Dreborg, K. (1996). Essence of backcasting. *Futures*, 28, 813–828. https://doi.org/10.1016/S0016-3287(96)00044-4.

Dubey, A., Naik, N., Parikh, D., Raskar, R., & Hidalgo, C.A. (2016). Deep learning the city: Quantifying urban perception at a global scale. In B. Leibe, J. Matas, N. Sebe, & M. Welling (Eds.), *Computer vision – ECCV 2016*. ECCV 2016. Lecture notes in computer science, vol. 9905. Springer. https://doi.org/10.1007/978-3-319-4644 8-0_12.

Dunkrel, A., Hartmann, M.C., Hauthal, E., Burghardt, D., & Purves, R.S. (2023). From sunrise to sunset: Exploring landscape preference through global reactions to ephemeral events captured in georeferenced social media. *PLoS ONE*, 18(2), e0280423. https://doi.org/10.1371/journal.pone.0280423.

Foster, K. (2016). Geodesign parsed: Placing it within the rubric of recognised design theories. *Landscape and Urban Planning*, 156, 92–100, ISSN 0169-2046.

Frost, A, Sturt, T., Kynvin, J, & Gallardo S.F. (2022). *Communicating with data visualisation: A practical guide*. Sage Publications.

Goodchild, M.F. (2007). Citizens as sensors: The world of volunteered geography. *GeoJournal*, 69, 211–221. https://doi.org/10.1007/s10708-007-9111-y.

Gottwald, S., Brenner, J., Albert, C., & Janssen, R. (2021). Integrating sense of place into participatory landscape planning: merging mapping surveys and geodesign workshops. *Landscape Research*, 46(8), 1041–1056. https://doi.org/10.1080/01426397.2021.1939288.

Gray, S., O'Brien, O., & Hügel, S. (2016). Collecting and visualising real-time urban data through city dashboards. *Built Environment*, 42(3), 498–509. https://www.jstor.org/stable/44132293.

Guillén-Mena, V., Quesada-Molina, F., Astudillo-Cordero, S., Lema, M., & Ortiz-Fernández, J. (2023). Lessons learned from a study based on the AHP method for the assessment of sustainability in neighborhoods. *MethodsX*, 11, 102440, ISSN 2215-0161. https://doi.org/10.1016/j.mex.2023.102440.

Harsia, E., & Nummi, P. (2024). Beyond the blind spot: Enhancing polyphony through city planning activism using public participation GIS, citizen participation, digital agency, and urban development. *Urban Planning*, ISSN: 2183-7635. https://doi.org/10.17645/up.7096.

Harvey, F. (2013). To volunteer or to contribute locational information? Towards truth in labeling for crowdsourced geographic information. In D. Sui, S. Elwood, & M. Goodchild (Eds.), *Crowdsourcing geographic knowledge*. Springer. https://doi.org/10.1007/978-94-007-4587-2_3.

Helbig, C., Becker, A.M., Masson, T., Mohamdeen, A., Sen, Ö.O., & Schlink, U. (2022). A game engine based application for visualising and analysing environmental spatiotemporal mobile sensor data in an urban context. *Frontiers in Environmental Science*, 10, 952725. https://doi.org/10.3389/fenvs.2022.952725.

Hoffman, J., Pelzer, P., Albert, L., Béneker, T., Hajer, M., & Mangnus, A. (2021). A futuring approach to teaching wicked problems. *Journal of Geography in Higher Education*, 45(4), 576–593. https://doi.org/10.1080/03098265.2020.1869923.

Hosseini, M., Miranda, F., Lin, J., & Silva, C. (2022). CitySurfaces: City-scale semantic segmentation of sidewalk materials. *ArXiv*. https://doi.org/10.1016/j.scs.2021.103630.

Hosseini, M., Sevtsuk, A., Miranda, F., Cesar Jr, R.M., & Silva, C.T. (2023). Mapping the walk: A scalable computer vision approach for generating sidewalk network datasets from aerial imagery. *Computers, Environment and Urban Systems*, 101 (2023) 101950.

Huang, L., Xiang, W., Wu, J., Traxler, C., & Huang, J. (2019). Integrating geodesign with landscape sustainability science. *Sustainability*, 11, 833. https://doi.org/10.3390/su11030833.

Huber, C., Nagel, T., & Stuckenschmidt, H.A. (2023). Initial visual analysis of German City Dashboards, The Eurographics Association. https://doi.org/10.2312/evp.20231056.

Huck, J., Whyatt, D., & Coulton, P. (2014). Spraycan: A PPGIS for capturing imprecise notions of place. *Applied Geography*, 55, 229–237.

Hudson-Smith, A. (2022). Incoming metaverses: Digital mirrors for urban planning. *Urban Planning*, 7(2), 343–354. https://doi.org/10.17645/up.v7i2.5193.

Lipson, M.J., Grimmond, S., Best, M., Abramowitz, G., Coutts, A., Tapper, N. et al. (2024) Evaluation of 30 urban land surface models in the Urban-PLUMBER project: Phase 1 results. *Quarterly Journal of the Royal Meteorological Society*, 150(758), 126–169. Available from: https://doi.org/10.1002/qj.4589.

Lovins, A. (1977). *Soft energy paths: Toward a durable peace*. Friends of the Earth/Ballinger.

Lynch, K. (1960). *Image of the city*. MIT Press.

Malleson, N. et al. (2022, January 1). Agent-based modelling for urban analytics: State of the art and challenges. AI Communications, 35(4), 393–406.

Mäntysalo, R., Granqvist, K., Duman, O., & Mladenović, M.N. (2023). From forecasts to scenarios in strategic city-regional land-use and transportation planning. *Regional Studies*, 57(4), 629–641. https://doi.org/10.1080/00343404.2022.2058699.

Marsal-Llacuna, M. (2020). The people's smart city dashboard (PSCD): Delivering on community-led governance with blockchain. *Technological Forecasting and Social Change*, 158, 120150. https://doi.org/10.1016/j.techfore.2020.120150.

Mattern, S. (2017). Urban dashboards. In R. Kitchin, T.P. Lauriault, & M.W. Wilson (Eds.), *Understanding spatial media*. Sage Publications. https://doi.org/10.4135/9781526425850.

Mattern, S. (2015, March). Mission control: A history of the urban dashboard. *Places Journal*. Accessed 06 Apr 2024. https://doi.org/10.22269/150309.

Milgram, P., & Kishino, F. (1994). A taxonomy of mixed reality visual displays. *IEICE Transactions on Information and Systems*, E77-D, 1321–1329.

Moffitt, L. (2023). *Architecture's model environments*. UCL Press. https://www.uclpress.co.uk/products/211136.

Moreira, G., Hosseini, M., Nipu, M. N., Lage, M., Ferreira, N., & Miranda, F. (2023). The urban toolkit: A grammar-based framework for urban visual analytics. *ArXiv*. https://doi.org/10.1109/TVCG.2023.3326598.

Müller, P., Wonka, P., Haegler, S., Ulmer, A., & Van Gool, L. (2006). Procedural modeling of buildings. In *ACM SIGGRAPH 2006 papers (SIGGRAPH '06)* (pp. 614–623). Association for Computing Machinery, New York, NY. https://doi.org/10.1145/1141911.1141931.

Parish, Y.I.H., & Müller, P. (2001). Procedural modeling of cities. In *Proceedings of the 28th annual conference on Computer graphics and interactive techniques (SIGGRAPH '01)* (pp. 301–308). Association for Computing Machinery, New York, NY. https://doi.org/10.1145/383259.383292.

ReMap Lima. (2014, February 6). *Point cloud of Lima*. https://remaplima.blogspot.com.

Rittel, H.W.J., & Webber, M.M. (1973). Dilemmas in a general theory of planning. *Policy Sciences*, 42(4), 155–169. https://doi.org/10.1007/BF01405730.

Robinson, J. (1982). Energy backcasting: A proposed method of policy analysis. *Energy Policy*, 10, 337–344. https://doi.org/10.1016/0301-4215(82)90048-9.

Roumpani, F. (2013). Developing classical and contemporary models in ESRI's CityEngine (CASA Working Paper No. 191). Centre for Advanced Spatial Analysis, UCL. https://www.ucl.ac.uk/bartlett/casa/publications/2013/may/casa-working-paper-191.

Roumpani, F. (2022). Procedural cities as active simulators for planning. *Urban Planning*, 7(2), 321–329. https://doi.org/10.17645/up.v7i2.5209.

Saldaña, M. (2015, December). An integrated approach to the procedural modeling of ancient cities and buildings. *Digital Scholarship in the Humanities*, 30(1), i148–i163. https://doi.org/10.1093/llc/fqv013.

Steinitz, S. (2012). *A framework for geodesign: Changing geography by design*. ESRI Press.

Sui, D., Elwood, S., & Goodchild, M. (2012). *Crowdsourcing geographic knowledge: volunteered geographic information (VGI) in theory and practice*. Springer Publishing Company, Incorporated.

Taillandier, P., Gaudou, B., Grignard, A., Huynh, Q.-N., Marilleau, N., Caillou, P., Philippon, D., & Drogoul, A. (2019). Building, composing and experimenting complex spatial models with the GAMA platform. *Geoinformatica*, 23(2), 299–322. https://doi.org/10.1007/s10707-018-00339-6.

Voros, J. (2003). A generic foresight process framework. *Foresight*, 5(3), 10–21. https://doi.org/10.1108/14636680310698379.

Voros, J. (2017). Big History and anticipation: Using Big History as a framework for global foresight. In R. Poli (Ed.), *Handbook of anticipation: Theoretical and applied aspects of the use of future in decision making*. Springer International. https://doi.org/10.1007/978-3-319-31737-3_95-1.

Wilson, A., & Tewdwr-Jones, M. (2020). Let's draw and talk about urban change: Deploying digital technology to encourage citizen participation in urban planning. *Environment and Planning B: Urban Analytics and City Science*, 47(9), 1588–1604. https://doi.org/10.1177/2399808319831290.

Wilson, A., & Tewdwr-Jones, M. (2021). *Digital participatory planning: Citizen engagement, democracy, and design* (1st ed.). Routledge. https://doi.org/10.4324/9781003190639.

Wu, J. (2019). Linking landscape, land system and design approaches to achieve sustainability. *Journal of Land Use Science*, 14(2), 173–189. https://doi.org/10.1080/1747423X.2019.1602677.

Yap, W., Stouffs, R., & Biljecki, F. (2023). Urbanity: Automated modelling and analysis of multidimensional networks in cities. *npj Urban Sustainability*, 3(45). https://doi.org/10.1038/s42949-023-00125.

Yap, W., & Biljecki, F. (2023). A global feature-rich network dataset of cities and dashboard for comprehensive urban analyses. *Scientific Data*, 10, 667. https://doi.org/10.1038/s41597-023-02578-1.

Young, G.W., & Kitchin, R. (2020). Creating design guidelines for building city dashboards from a user's perspectives. *International Journal of Human-Computer Studies*, 140, 102429. https://doi.org/10.1016/j.ijhcs.2020.102429.

CHAPTER 5

GAMING, WORLD-BUILDING, AND PARTICIPATORY PLANNING

INTRODUCTION

Digital games have long concerned themselves with abstracted simulations of our physical environment and continue a long tradition of strategy and foresight (Bradfield et al., 2005). In the case of war gaming there are precedents from a 'realistic' Prussian prototype gaming table and cabinet by George Leopold von Reisswitz (1812) which was further developed by Georg Heinrich Rudolf Johann von Reisswitz (1834) using topographical maps. The game continues to be played via the International Kriegsspiel Society today, exploring military scenarios, humanitarian aid, and policy. Tabletop games and digital games have been used in relation to scenario planning for plausible futures, particularly in urban environments. A notable example was the RAND Corporation's Mathematical Analytics Division's (MAD) development of physical games for social science to test military activity, design variables, play assumptions, and anticipate effects (Kahn, 1967; Daye, 2019) (Chapter 3). Games represent reality, and the abstractions they make are of particular interest, given UDT systems seek to assess federated models in order to 'twin' reality. In addition, through the act of playing games, we generate data on how we interact, respond, and make decisions. The act of digital gameplay, therefore, reveals fundamental data in the socio-technical use of UDTs. These interactions between games and planning could arguably be called a 'matter of concern' in Bruno Latour's terms, which consists of the study of entanglements, complexity, and the socio-technical relations between humans and objects (2004). What can digital games add to the urban planning space in terms of game systems, the simulation of complex urban issues, and how can game mechanics be deployed?

A working hypothesis developed by myself and Paul Coulton suggests that the range of city information models (CIMs) and UDTs utilise game technologies, particularly game engines, for interactions. However, they do not necessarily incorporate digital game mechanics and dynamics (Hunicke et al., 2004), levels, progress, flows, and feedback as part of world-building, and this needs to be more fully understood if such systems are to yield potential benefits in terms of citizen engagement. Design fiction has been discussed in Chapter 1 in terms of world-building approaches, which can offer alternatives and it is important to understand the mechanics of city-building games in urban planning, architecture, and design. This chapter develops this relationship through the development of a playable urban game continuum to illuminate the various nuances of a range of precedents and scaffold future applications. The interactive game continuum established here is supported by the establishment of a taxonomy of city-building games and real-world UDTs. The hypothesis of the use of game design contexts for improved UDTs is tested via an experimental case study of planning policy using imported geodata into the digital game *Cities: Skylines* with a large number of modifications (Colossal Order, 2015). The gameplay allowed a range of urban analyses, and the models created in the game are then compared to real-world strategic plans, play-tested by n.140 children (ages 7 to 13) and adults. The analysis utilised AI tools to create a report on the design 'intention' of children for the real-world space and create a feedback loop. The results of this activity help scaffold and locate usage of the urban game continuum interactive tool and explore 'game-like' aspects of planning simulation that can inform future UDTs.

WORLD-BUILDING GAMES AND SIMULATION TOOLS

Various terms circulate concerning digital and analogue games. Terms such as 'serious games' were conceptualised as a way to describe games aimed at simulating aspects of life in procedural and rationalised manners for education and professional training while engaging in the aspects of 'play,' with the term emerging from Clark C. Abt (1970). 'Serious' games have emerged as an approach often utilised for urban planning (Djaouti, 2011). In terms of digital games

CHAPTER 5
GAMING, WORLD-BUILDING, AND PARTICIPATORY PLANNING

Figure 5.1
Mabel Addis, 1964. The Sumerian Game (IBM 7090). BOCES; IBM. Photograph of 6th Grader, Joanne Chomich playing the IBM terminal, published in Computerworld Volume 2, Issue 14 on April 3, 1968. Open access following the work of The McGovern Foundation.

for urban planning, the first work emerged from the designer and programmer Mabel Addis Mergardt and William McKay who created *The Sumerian Game* (Addis, 1964), an IBM mainframe planning game written in Fortan, designed for economic instruction for school children, which was set in Lagash in Sumer circa 3500 BC (Figure 5.1). The Summerian game was a text-based game in which players managed various rounds (seasons) of land management with projections and an audio guide for each round, giving procedural choices for players. Doug Dyment later recreated the digital game in (1968) under the name Hamurabi (Dyment, 1968) (this was earlier titled *King of Sumeria* or *The Sumer Game*). The game inspired various other versions, including George Bank, *Santa Paravia en Fiumaccio* (Blank, 1978) in which a player ruled an Italian city-state in 1400 through turn-based moves and city-building capability (Figure 5.2).

219

Figure 5.2
George Blank, 1978. Santa Paravia en Fiumaccio. Instant Software.

These early digital game examples helped create a trajectory for city-building simulators, which have emerged as a creative tool for exploring urban governance, city simulation, and planning. Of this trajectory, of note a two-player game by Don Daglow, Utopia (Daglow, 1982), featured various construction options as a precursor to real-time strategy games (Daglow, 2018). Arguably, the most famous educational city-building game in this timeline is SimCity (Maxis, 1989), also known as *Micropolis* (Figure 5.3). Will Wright formed Maxis with businessman Jeff Braun to self-publish *SimCity* in 1989. *Micropolis* allows the players to inhabit the character of the city mayor through an isometric top-down view to develop various planning models, forms, and city structures, particularly through the use of zoning (Gaber, 2007; Terzano & Morckel, 2017). *SimCity* used a form of agent-based modelling (ABM), cellular

CHAPTER 5
GAMING, WORLD-BUILDING, AND PARTICIPATORY PLANNING

Figure 5.3
Don Hopkins, 1989, Micropolis, Kowloon City, Maxis, Electronic Arts Inc. Micropolis on OpenLaszlo/Flash Web Client and Python/C++ Web Server. https://www.donhopkins.com/home/micropolis/ & https://github.com/SimHacker/micropolis.

automata (CA) modelling, which involves 'agents' and interactions between things in order to model complexity, sometimes through the use of infinite cell structures and relationships (see chapter case study) (Crooks et al., 2021; Yeh et al., 2021). ABM and CA are processes that share system rules similar to those of city-building games but allow much more developed simulation and analysis. *SimCity* is CA with agents, though the design motivation is very different. From a user-centric perspective, playing SimCity can explore various scenarios, and this is termed 'black box' testing where the internal system rules are unknown. White box testing, however, examines the specific internal code of the game and structure, analysing components and functions. Black and White box testing in combination are used interchangeably to produce an overall qualitative test. Yet, how are black and white box testing applied to UDTs? What are the underlying parameters compared to city-building games? For *Micropolis* (*SimCity*), as expanded by Wright, the focus is on the fiction of a city for play.

221

> *(...) you give the player a tool so that they can create things. And then you give them some context for that creation. You know, what is it, what kind of kind of world does it live in, what's its purpose? What are you trying to do with this thing that you're creating? To really put the player in the design role. And the actual world is reactive to their design... Giving them a pretty large solution space to solve the problem within the game. So the game represents this problem landscape.*
>
> (Pearce, 2022)

Cities and urban areas are complex systems, and the use of games allows a player to explore the complexities of such systems, to which Wright is referring, and simulate and model behaviour and realise scenarios. Fundamentally, a broad range of stakeholders is required to be embedded in the creation of UDTs because this situates people in that very problem space to which geodesign comes to the fore. However, city-building games are designed to provide pleasurable experiences and are not planning simulations, there is a difference in the intention (Devisch, 2008, Bogost, 2011). In the game SimNimby (Nass & Weeks, 2022) (Figure 5.4), which is a subversion of Micropolis, players progress through futile planning experiences encompassing 54 different development objections, which eventually end the game and is based on the real-life experiences of the game designers working in San Francisco and Brooklyn, USA. Is *SimNimby* more like a real-world planning system?

This satire highlights some of the complexities experienced in real-world planning consultations. *SimCity* and the games showcased in this chapter introduce players to an abstraction of complex urban systems, often with little adherence to real-world planning rules and procedures. SimNimby is a representation of a real-world planning system, which is presented as a parody to highlight the

CHAPTER 5
GAMING, WORLD-BUILDING, AND PARTICIPATORY PLANNING

Figure 5.4
Steven Nass. Owen Weeks, 2022, SIM Nimby. https://stevenjnass.com/.

frustration that often occurs when engaging with such technocratic schemes. As a spin-off of Maxis, Maxius Business Simulations created various corporate SIMs, including the prototype SimRefinery (Maxis, 1992) for Chevron, to simulate complex real-world oil refineries (Figure 5.5). It should also be noted that *SimCity* and others follow the dominant paradigms of planning systems and do not feature alternatives in their black box. For example, Bradley Bereitschaft (2016) defines *SimCity* as 'gentrification' and represents car-based preferential futures.

Games can also influence real-world policy. The release of SimHealth (Thinking Tools, 1994) sought to simulate and game American healthcare during the Clinton healthcare plan reforms, juxtaposing the simulation with real-world health politics. The Markle Foundation commissioned Maxis Business Solutions to create *SimHealth* in order to provide a game experience of complex policy and health care. The game was released on Capitol Hill and copies were provided to lobbyists and the White House (Salvador, 2020). Reflecting on *SimHealth* at the time of release, and his role as a policy maker, Paul Starr, states,

223

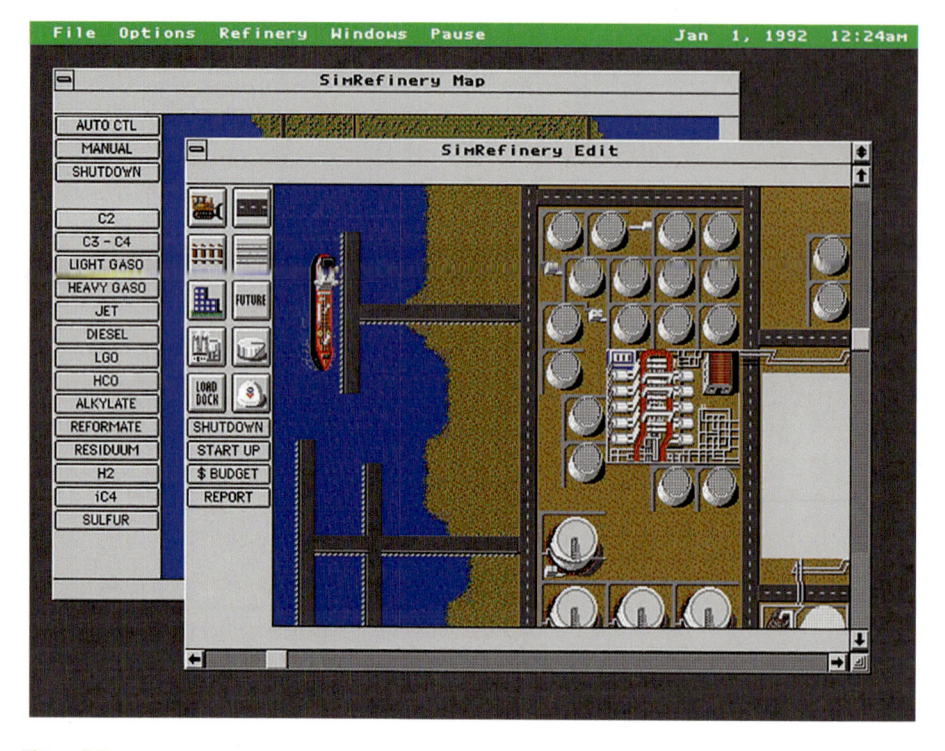

Figure 5.5
Will Wright and Jeff Braun, John Hiles, Prototype of Sim Refinery by Maxis for Chevron, 1993, Maxis Business Simulations, Chevron Corporation.

> *(...) when policymakers depend on simulations to guide present choices–especially when legislators put government on "automatic pilot," binding policy to numerical indicators of projected trends–they cede power to those who define the models that generate the forecasts.*
> (Starr, 1994, p. 20)

Starr highlights an important research space for using gaming technology and mechanics in urban planning platforms and systems.

Digital games embed particular geographic imaginaries and concepts through the act of world-building and strategy. For

Figure 5.6
Celestial Software. (1991). Utopia: The Creation of a Nation. Gremlin Interactive. Designers Graeme Ing and Robert Crack. Image courtesy of Ian Stewart.

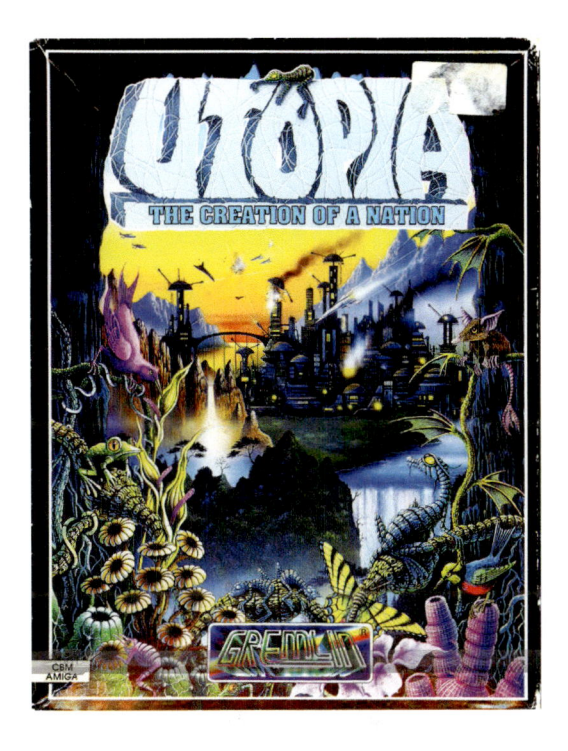

instance, *Utopia: The Creation of a* Nation (Celestial Software, 1991) is a micromanagement simulation game on an off-world colony inspired by *SimCity* (Figure 5.6). In Utopia, players compete against alien races to establish colonies in ten scenarios on an isometric playboard with a highly considered user interface to which design lessons have a crossover to city dashboards. Additionally, it included a disaster scenario expansion pack with various porting options.

One of the most recent city-building games, Cities XL (Focus Entertainment, 2013), allowed the player community to modify the game to build real-life cities such as Tokyo (Figure 5.7). In Colossal Order's Cities: Skylines (2015, version II, 2023), community modification is further increased with players modelling brands, housing technologies, and transport in high-resolution detail, reflecting the wide variety of global urbanisation.

Figure 5.7
Cities XL, Focus Entertainment, 2013, Monte Cristo; Pullup Entertainment. Authors in-game screen.

This city-building genre maintains itself as a core game genre and meaningful game experiences, such as Block'hood (Plethora Project, 2016) Figure 5.8, in which players must envision a neighbourhood and maintain ecological balance, provides a real-world planning problem space; Constructor (System 3, 1997) in which players inhabitant a commercial construction company and undermine its rivals as an example of commercially driven urbanism, and *Little Cities* (nDreams, Purple Yonder, 2022) (Figure 5.9) in which players develop on grid forms on an island and the prosperity is measured through the immersive design choices. Finally, *The Architect: Paris* (5PM Studio, 2021) is a high-realism city planning sandbox based in Paris, France, with extensive urban design block and architectural detailing features in which players can realise designs as an architectural firm (Figure 5.10). This timeline of city-building games has been used to inform the taxonomy of games and UDTs.

CHAPTER 5
GAMING, WORLD-BUILDING, AND PARTICIPATORY PLANNING

Figure 5.8
Plethora Project, 2016, Block'hood, Devolver Digital.

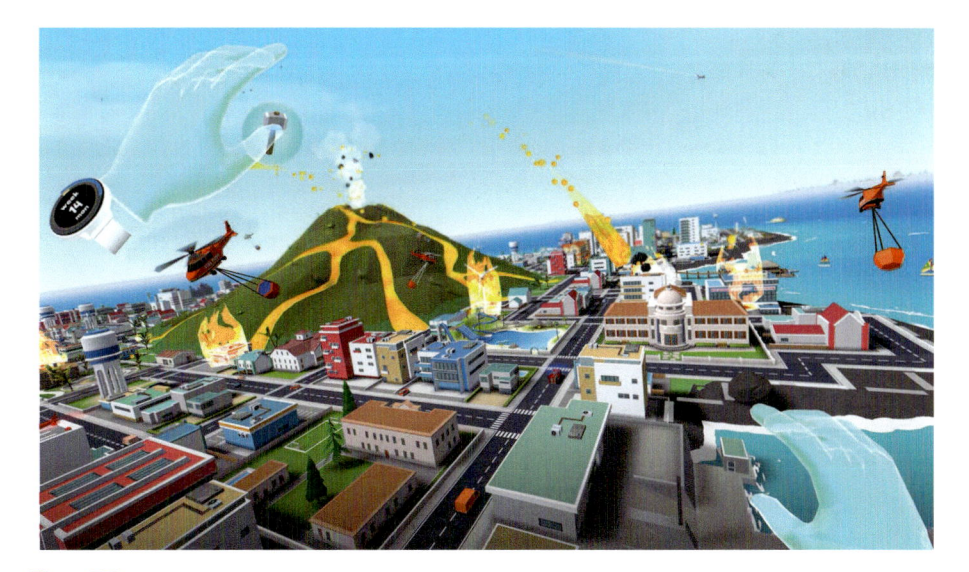

Figure 5.9
Little Cities VR, 2024, nDreams Limited.

227

Figure 5.10
The Architect: Paris, 2021, Enodo Games Inc., Focus Entertainment. Oscar Maidment and Paul Cureton in-game screenshots.

GAMING PLANNING SYSTEMS

If you, the reader, were a games designer tasked with creating a digital game of a national planning framework and tasked with creating local planning policy to play, the complexity of this world-building exercise would be highly challenging. For example, in the United Kingdom, The National Planning Policy Framework (NPPF) is the UK government's view of the planning system in England that must operate in accordance with primary and secondary legislation with 16 categories under the framework provision (Figure 5.11). The NPPF provisions are supported by further detail in the National Planning Practice Guidance (NPPG). Aligned with the NPPF, 317 local planning authorities (LPAs) develop local policies, plans, and neighbourhood plans. Updates to the NPPF are frequent, often yearly. Thus, creating a simulation of the NPPF and NPPG is not conducive to game design principles, nor can the nuances and translation of a framework, primary and secondary legislation, and a range of local plans into game rules be properly enacted in totality.

Figure 5.11
National Planning Policy Framework (NPPF) in the United Kingdom. Authour diagram based on the "Department for Levelling Up, Housing & Communities, Levelling-up and Regeneration Bill: Reforms to National Planning Policy" (2023).

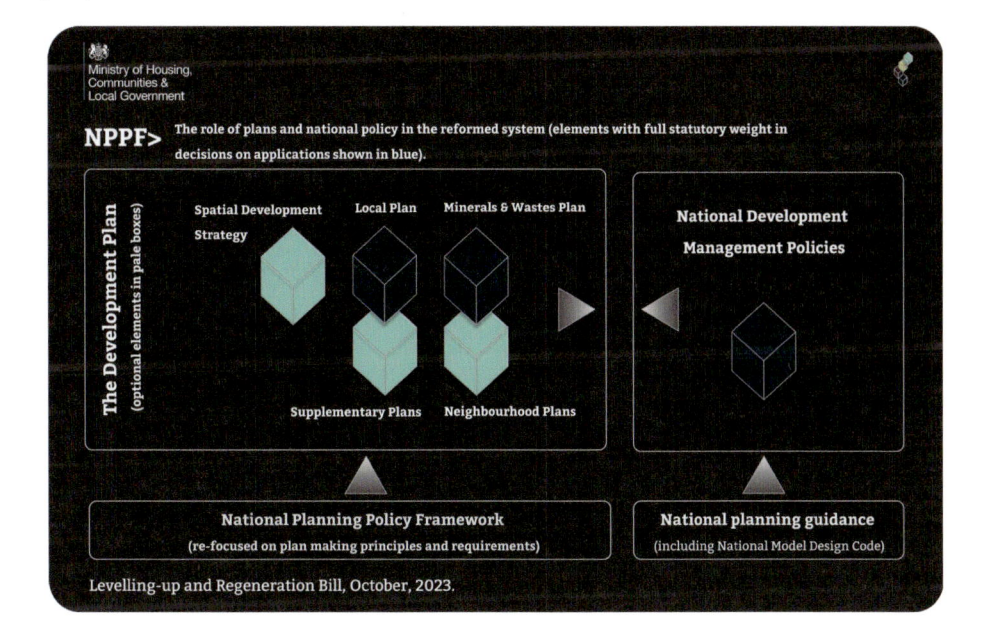

However, this book has charted the rise of city information models (CIMs), a term coined by Lachmi Khemlani in 2005, 3D geospatial data, cloud-based services for the web, and urban digital twins (UDTs). Developers have started to incorporate various game technologies and design patterns to create interactive systems around particular aspects of the planning system such as land release platforms and 3D visualisations of major developments as well a climate atlas (*Energia- ja ilmastoatlas* [Energy and Climate Atlas], (n.d.). There is an extensive crossover to account for in terms of UDT system elements, particularly 3D GIS, urban informatics, data dashboards and XR in terms of walkthroughs, client demonstrations, and public consultations. There is an opportunity to define the gamified and game-like systems being developed in relation to actual city-building games and game design principles in order to both understand current practices and open up the opportunity to playfully engage the general population in the complexities affecting decision-making in urban planning. The opportunities afforded by digital games also address the diversity of stakeholders and scenario-generation possibilities as part of the geodesign process.

Currently, many of the UDTs seek participation using gaming technology for aspects of their systems, yet miss the game's design elements, which could provide much more positive experiences and the broadening of the stakeholders involved in the creation of UDTs. For example, Markus Persson and Jens Bergensten's *Cave Man*, which subsequently was renamed Minecraft (Mojang Studios, 2011), is an open-world explorative free block creation game and has been translated into the UN-Habitat *Block-by*-Block methodology from 2012, as we have seen in Chapter 3 (Imam & Lahoud, 2021), and the use of Minecraft for geophysical properties and sub-surface representation (Conclusion). Communities use *Minecraft* to plan, design, and test proposals for public spaces using the game. This example is an outlier compared to current systems and cases focused on the exploitation of gaming technology only and there are opportunities to engage players in changing the rules of the system being replicated.

The use of digital games is much needed as governments such as the UK argue that urgent change is required in planning to address the chronic shortfall in housing and stimulate sustainable economic growth. Digital games enable the playing of our urban futures. Gaming technology and gamified geodata are intended for citizen participation and access, yet fundamental challenges remain unaddressed. The Royal Town Planning Institute (RTPI) in the UK stated that "response rates to a typical pre-planning consultation are around 3% of those directly made aware of it. In local plan consultations, this figure can fall to less than 1% of the population of a district" (Manns, 2017). The RTPI paper of 2020 also evidenced that only 11% of young people have been engaged in a local plan consultation (Butler et al., 2020, p. 29). However, what if games research and ludic interactions are introduced and gaming technology in which the planning system is played and redesigned? This potential would create alternative ways of working and new possibilities for world-building (as found in design fiction) developing fundamental changes to the planning system. Games should not be vehicles in which real-world planning systems are simulated, but the situation should be reversed, and digital urban games should be a platform in which alternative worlds are imagined in order to change the limitations of that very real-world system. Would this world-building approach and result increase citizen participation, and what would this game system for planning look like?

URBAN GAME CONTINUUM

Utilising the range of city-building game precedents mentioned earlier in this chapter, an urban game continuum is presented in Figure 5.12. This interactive tool isolates four major factors in this chapter. The continuum has four level blocks. The continuum operates left to right in a similar way to the virtuality continuum created by Paul Milgram and Fumio Kishino, often used as an initial basis to discuss extended reality (XR) and the metaverse (Milgram et al., 1995).

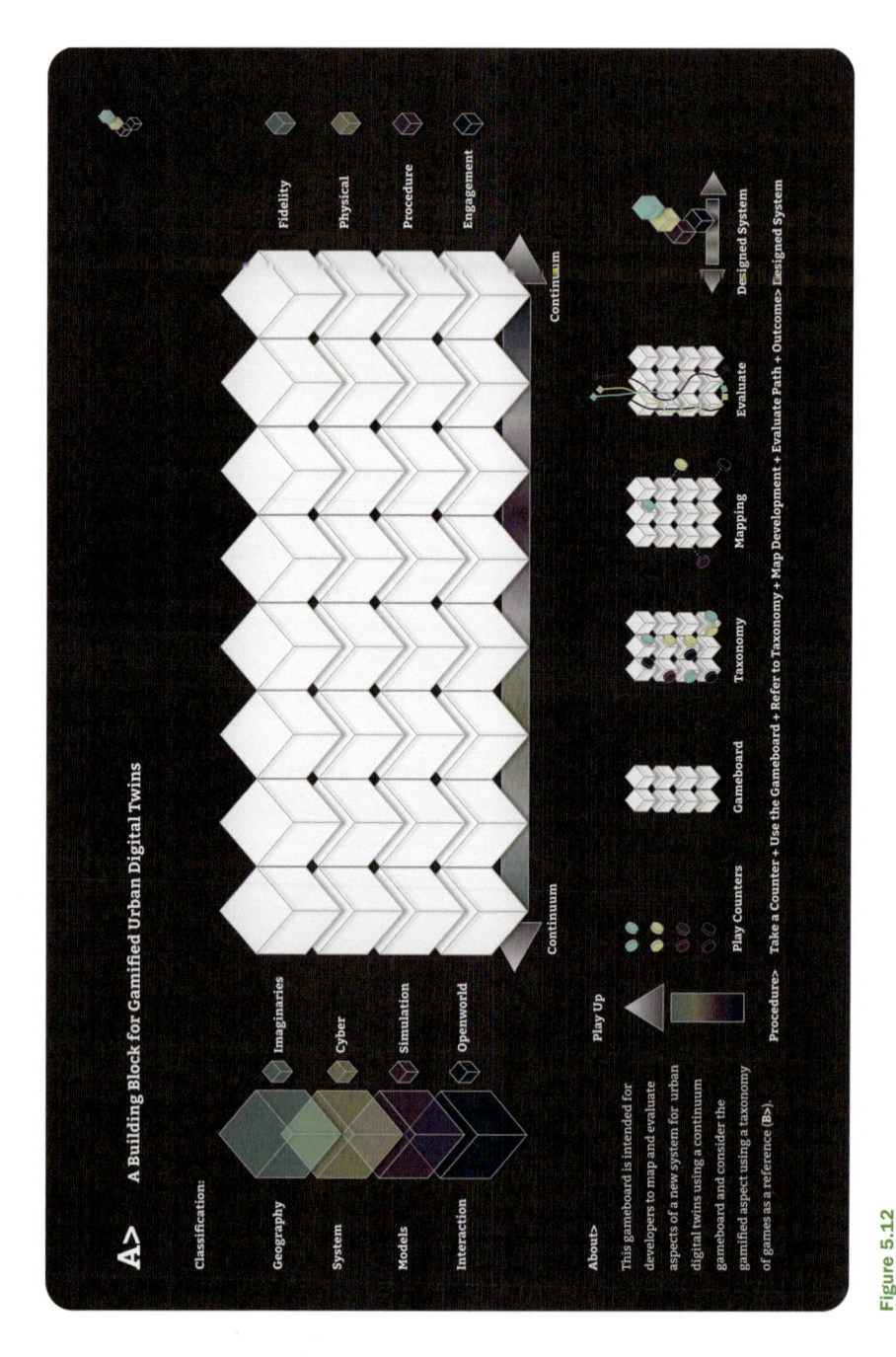

Figure 5.12
Authour, Urban Game Continuum, Urban Game Continuum Interactive Tool and Taxonomy, 2023. Figma. Available at: https://www.figma.com/community/file/1374999663066720955.

The interactive tool contains four classifications in the coloured level blocks – geography, system, models, and interactions which are common attributes of digital games, 3D GIS, CIMs, and UDTs. Players map from the bottom upwards, plotting and mapping the aspects of their own future system using coloured counters. The resultant map thus informs the system design. The four classifications are devised on these terms:

1. Interactions: Acts of open-world gaming (left), single-player, ludic and games technology for citizen engagement (right).
2. In terms of models and urban simulations, the continuum maps simulations of implausible acts and environments (left) and augmentation of real-world urban planning (right).
3. In system terms, a continuum for cyber networks, game systems (black box), data and models (left) and their relationship with the physical (white box, cyber-physical) is created (right).
4. In geographic terms, a classification and continuum are created to map pure imaginaries (left) and games and UDTs that adopt high realism and fidelity (right).

To map a new system onto the continuum and on the four classifications, a supporting taxonomy, which is hyperlinked to the original game, case, or dataset, is developed from city-building games, 3D GIS, CIMs, and UDTs to support this exercise. These two components allow users/players to reference and benchmark the games and cases against their development needs. For example, a player may design a system on the continuum from left to right and benchmark from the taxonomy. For example, a user/player mapping a new system may plot their design by comparative values such as one more block face for open-world exploration compared to the SimCity 2000 game, two more block faces towards *Procedure* nearer the game *Cities: Skylines*, and one more block face towards *Physical*, and three more block faces across to *Fidelity*. Establishing

GEODESIGN, URBAN DIGITAL TWINS AND FUTURES

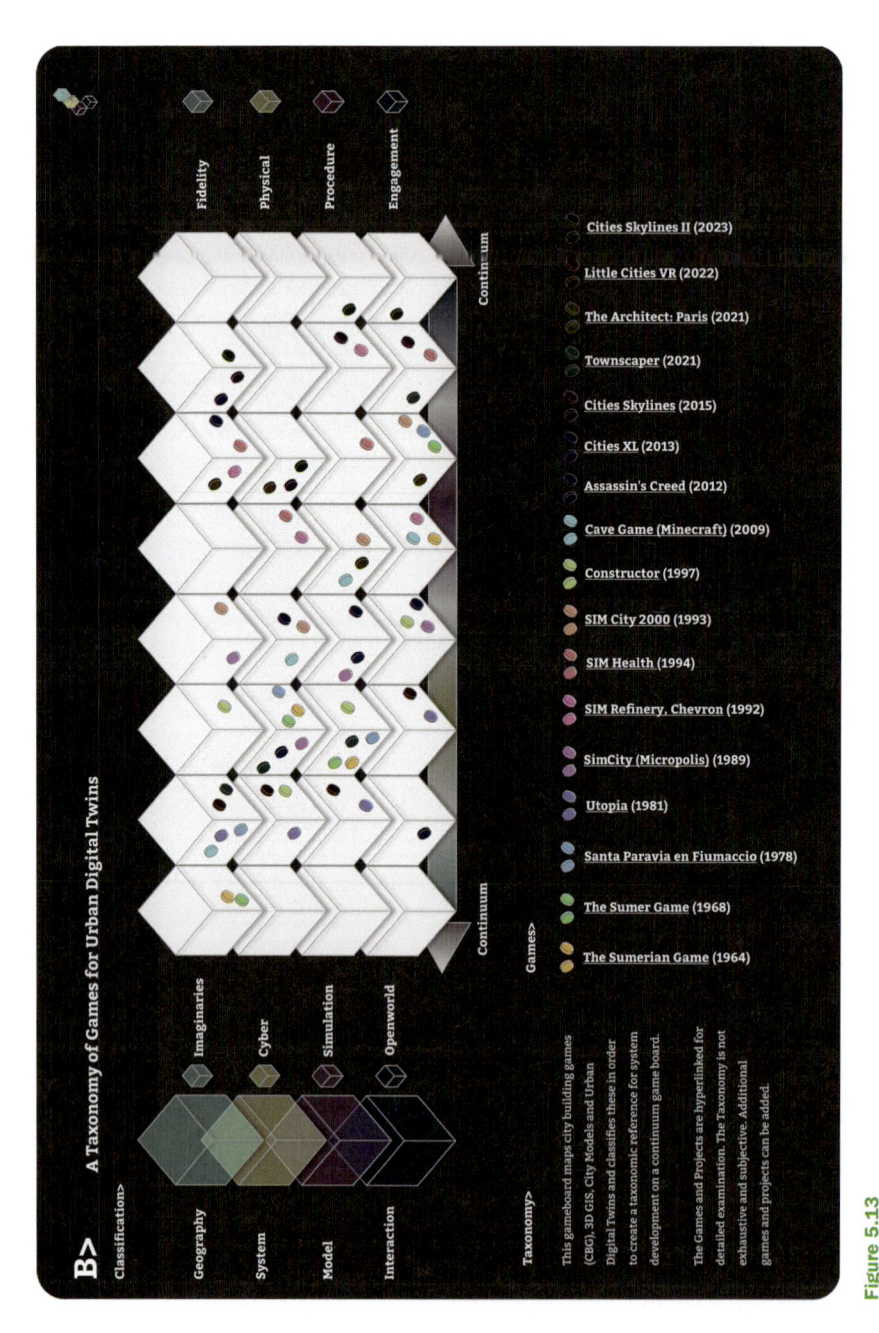

Figure 5.13
Authour, Taxonomy of City Building Games and Urban Digital twins, Urban Game Continuum Interactive Tool and Taxonomy, 2024. Figma. Available at: https://www.figma.com/community/file/1374999663066720955.

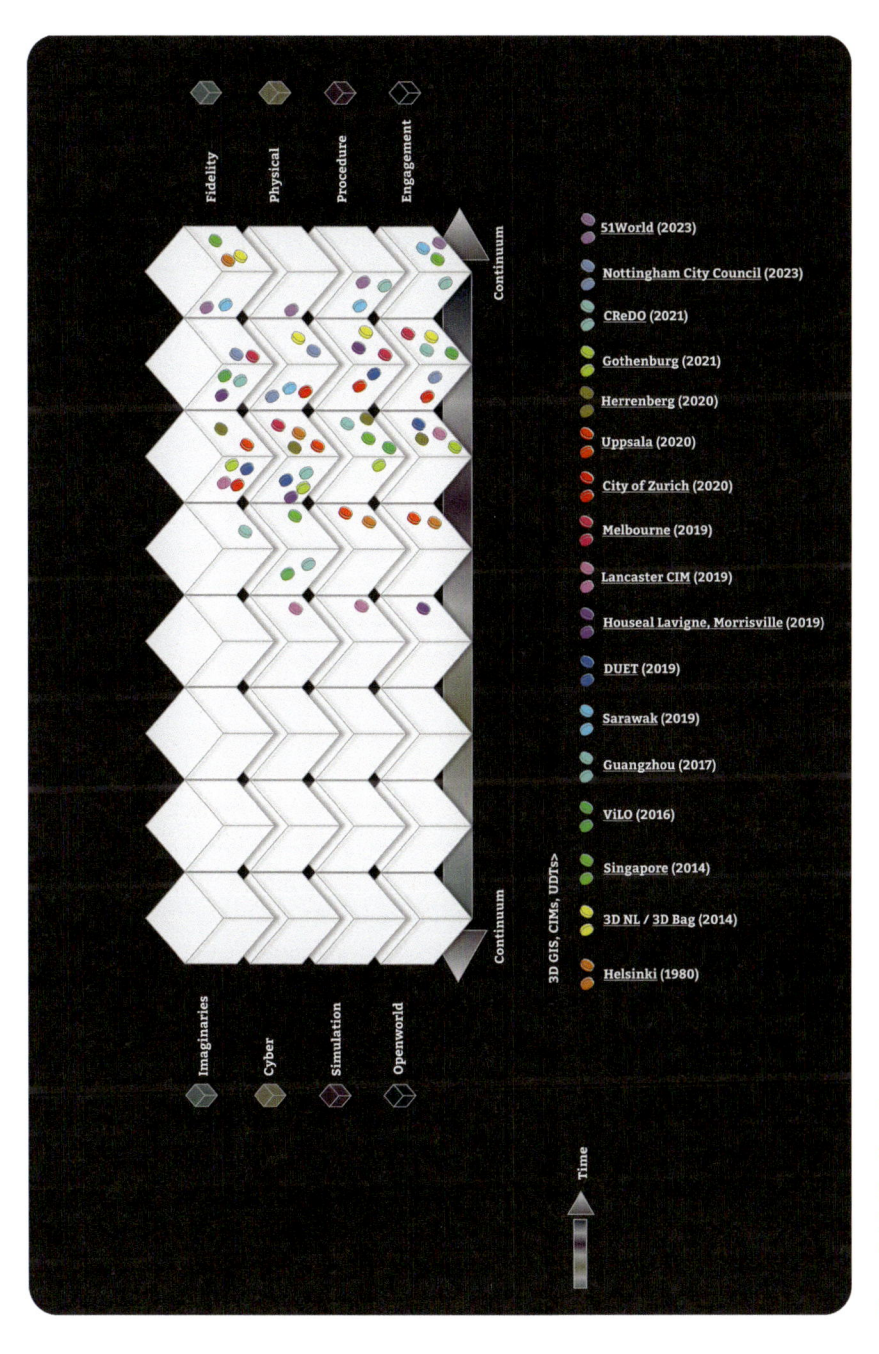

CHAPTER 5
GAMING, WORLD-BUILDING, AND PARTICIPATORY PLANNING

Figure 5.13 (Continued)

235

such a continuum is essential as there is a rich territory and crossover with the increasing urbanisation and emergent possibilities of UDTs.

Supporting the design of this continuum is a taxonomic reference that is subjective, not exhaustive, and expandable from 17 games and 17 UDTs plotted in the first instance (Figure 5.13). In order to create this continuum, distinctions need to be made between the use of games for planning education and game technology for planning processes. 'Game-like' platforms, such as Cesium plugins for Unreal Engine and ESRI ArcGIS, provide 'real-world' geolocated information but do not provide a 'playful' experience. In an example balancing games experiences and real-world planning, revisiting the example from Chapter 2, Houseal Lavigne used a procedural modelling program and a games engine as part of public consultation for Morrisville, North Carolina, USA (2019) and its land use plan in which two scenarios and key performance indicators were created for its town centre with different build-outs, including a dashboard that adopted the zoning code ("Mapping Morrisville," 2021). Hence, it is relatively central on the gameboard, leaning towards fidelity and engagement (Figure 5.14).

Game engines are utilised for a range of activities and simulations of synthetic data, including geolocated weather and 3rd-person open-world exploration, amongst other operations. GIS and game-engine integration is, of course, reliant on streaming architecture. The gamification of GIS systems, availability of photogrammetry data, satellite imagery, and open street map provide environments long sought in early city-building games. Gaming technology enables fidelity and urban planning simulations to a higher degree (Poplin et al., 2020). Thus, in the UDT 17 cases, the majority are mapped towards the right-hand side of the continuum and support the hypothesis of the one directionality of games technology applied towards urban planning, with many lacking game design principles, though there are outliers such as Houseal Lavigne which have achieved a careful balance.

CHAPTER 5
GAMING, WORLD-BUILDING, AND PARTICIPATORY PLANNING

Figure 5.14
Town of Morrisville and Houseal Lavigne, Mapping Morrisville, 2021. Available via ArcGIS Storymaps. https://storymaps.arcgis.com/stories/4cdd4c970be8451c8de734532ed6e2d1.

The role of digital games is critical; for example, a range of planning platforms utilising 3D GIS and data dashboards to simulate future scenarios and describe developments in the current planning process. In comparison with the digital game Cities: Skylines II, there is a user interface crossover with each respective project using key urban indicators. In both cases, there is a need for navigable, understandable, and progressive information in order to explore the various parameters of the fictional or real-world 'gamified' space (Young & Kitchin, 2020).

237

The crossovers of user interfaces, data visualisation, and 3D models are apparent in all cases of the taxonomy. The mechanics also crossover, such as navigation controls, first-person immersion, and simulated weather, such as the use of virtual sun shading analysis or line-of-sight studies. However, unlike city-building games with pre-defined missions, progression, and open-world creativity, this 'gamified' content has profound implications. Does the black box space of the game-like UDT platform and model dictate decisions, and how does this change the dynamic of the planning process and progression judgements? In other words, games like UDTs are played out. The social-technical relations between systems and interaction are thus a critically underexplored research space. Thus, games and the range of real-world planning systems using game technology are critical research areas.

MODIFYING CITIES: SKYLINES WITH GEODATA, LANCASTER CITY, UK

Considering the urban game continuum and taxonomy presented in this chapter, the relationship between digital games, planning, and UDTs as part of geodesign is further explored through experimentation. In a digital game *Cities: Skylines*, modification and map builder are utilised to import real-world geodata to examine the game's simulative and reductive qualities and consider real-world planning issues (Khan & Zhao, 2021). This activity was designed so that it can be replicated at a low cost around the world. The purpose of the task is not to critique current local plans but to simulate some of the planning policy principles and explore options for world-building outside of the NPPF system in the UK, though this can also apply to other national planning systems. For the game *Cities: Skylines*, players can import geodata height maps (digital terrain model) and export fictional cities and models to real-world digital open street maps (OSM) via modifications or import real-world maps. The modifications to do this in *Cities: Skylines II* (Colossal Order, 2023) were not yet available at the time of writing. There is already an established workflow, though there are opportunities to demonstrate the bi-directionality of the gaming platform itself (Wicaksana & Darmawan, 2022). A range of precedents exist showcasing this workflow including, Norra Djurgårdsstaden, Stockholm, Sweden (2017), the role of industrial zones and factories in Braunschweig, Germany (2017) (Juraschek et al., 2017), Żuromin, Central Poland (Olszewski et al., 2020), and Denpasar City, Indonesia (2021), amongst many others. The game can also simulate historical plans and paradigm city structures, such as Ebenezer Howard's layout for Letchworth Garden City (2022) (City Planner Place, 2022). Realistic 1:1 scale cities, such as Rochester and Pittsburgh, USA, have been realised, demonstrating possibilities with additional build hours and skills (Caseous Stranger, 2022). This translatory aspect of the game towards real-world planning issues has also been applied, allowing game players to understand sustainable development goals (SDGs) in their game world-building (Jolly & Budke, 2023). The extent and limitations of the game can be mapped to the fidelity, physical replica, procedures, and engagement values from the urban game continuum and are mapped based on the game mechanics.

This chapter's case is mapped against the degrees of imaginaries and fidelity of the game in terms of managing health, employment, traffic, and pollution levels compared to the real-world strategic planning policies of Lancaster City Council in Lancaster, UK. The degree to which *Cities: Skylines* can relay the cyber or physical aspect of the city does have limits as a simulation, the game 'black box' capacity has to be understood in its ability to mimic planning procedures. However, the engagement potential of the game and playful experience in order to navigate real-world issues through game modifications holds much potential. Indeed, much of the explorations focus on one aspect of the game, real-world planning issues, rather than the possibilities of new modifications via gaming and redesign of the existing real-world planning system through a world-building approach.

A large section of the Lancashire district region has been used in the game (Figure 5.15), North-West UK (Lng/-2.7984740523877747 Lat/54.044485589053494). The district has a population of 144,246 and area coverage of 576.2 km²/57,620 ha. The area was chosen for its spatial complexity, and the non-metropolitan district is defined through UK Governmental Boundaries, its legislative framework, and the Office for National Statistics ONS ("Adopted policies," 2020a). The district is a two-tier non-metropolitan area of high-grade agricultural land with extensive rural coverage and mixes small towns, coastal communities, and post-industrial ports. The primary urban area is an arc stretching from the coast at Heysham, upwards to Morecambe Bay and southeast to Lancaster. The study area contains the Forest of Bowland, an Area of Outstanding Natural Beauty (AONB), Arnside & Silverdale (AONB), military sites, and a nuclear power plant. The

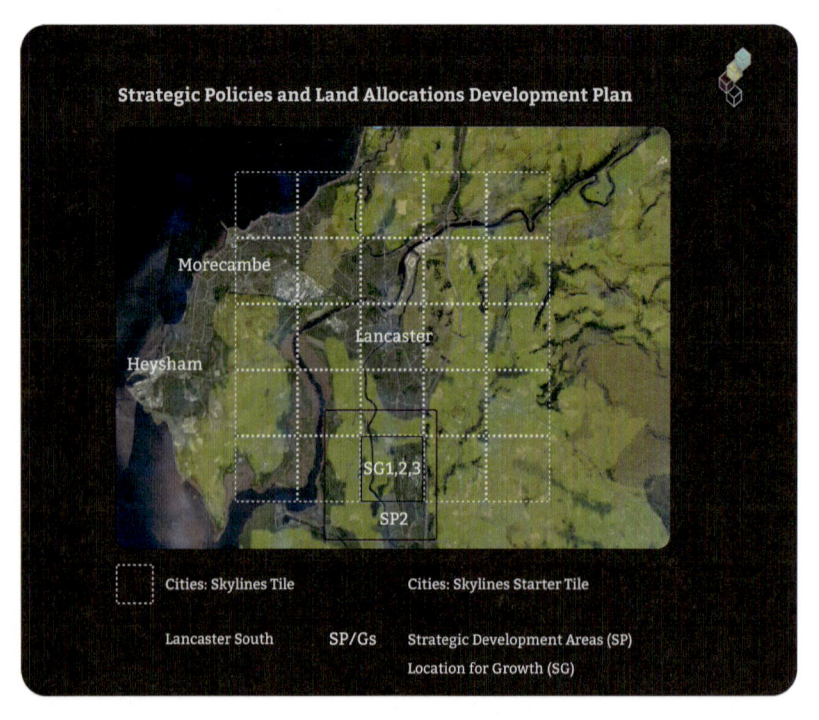

Figure 5.15
Authour Lancaster Strategic Policies and Land Allocation Development Plan & Cities: Skylines Map Tiles. Available at: https://steamcommunity.com/sharedfiles/filedetails/?id=3132777655&searchtext=lancaster.

CHAPTER 5
GAMING, WORLD-BUILDING, AND PARTICIPATORY PLANNING

current local plan for Lancaster District was adopted for 2011–2031 and contains the policies for sustainable settlements, rural villages, and future growth development areas (Lancaster City Council, 2020a).

The start game area of the map was chosen for Lancaster University, which is nestled between the M6 and West Coast mainline railway and Lancaster Canal, which is central in the 'Lancaster South Development Area' (Figure 5.16) (Lancaster City Council, 2020b, p. 51). *Cities: Skylines* requires the starting tile area, which was used for the detailed build-out south of the University, which is a sustainable settlement policy growth and employment area (SP2, SG2) in the adopted local plan at Galgate (Lancaster City Council, 2020b, p. 25). A planned broad settlement called Bailrigg Garden Village (SG1, SG3), delivered by Homes England to the west of the campus, has had outline planning for around 35,000 homes, and a junction link to the M6 road has been placed on hold. The planning policy for the garden village seeks community involvement in the development, of which *Cities: Skylines* could play a role (Bailrigg Garden Village, 2022). The authors used the policy principles and made reference to them during gameplay, as well as tested deviations from the policy. For example, for policy SG1, the following principles were considered in the gameplay guide:

- **The delivery of access into the Strategic Highways Network via a reconfiguration of Junction 33 of the M6 to the satisfaction of the strategic and local highways authority.**
- **Improvements to the local road network as appropriate to address recognised capacity issues and issues of highway safety to the satisfaction of the local highways authority.**
- **Improvements to the public transport network, specifically the creation of a Bus Rapid Transit System linking South Lancaster to Lancaster City Centre, Morecambe and the employment areas on the Heysham Peninsula to provide genuinely realistic alternatives to private vehicle use.**

241

Figure 5.16
Lancaster South Map & Cities: Skylines in-Game Play. Source: authors' screenshots from Cities: Skylines (Colossal Order, 2015).

- **Improved cycling and walking linkages from South Lancaster to the north, towards Lancaster City Centre and to the south, towards Galgate. This will be through the creation of a Cycling and Walking Superhighway which will provide a safe and attractive route for pedestrians and cyclists. Improvements will also be sought for improvements to walking and cycling links along the Lancaster Canal.**
- **The delivery of sufficient education places at both a primary and secondary school level to the satisfaction of the local education authority.**
- **The delivery of new local centre(s) provision, which will include a range of local services and community facilities in an accessible location for both new and existing residents in South Lancaster.**
- **The provision of sufficient public open space to fully meet the amenity and recreational needs of the residents in the Garden Village (Bailrigg Garden Village, 2022, p. 60).**

The content management system and indicators of the game provide clear progression and elements to build out the university campus and growth area. The game mechanics rest on building out from road networks and parking lots over pedestrian routes, and several city elements have to be incorporated, such as a fire station and hospital, in order to stimulate growth. There was often a divergence between the real-world development plans and the necessary components for gameplay. Transport networks often must be looped and do not reflect the real-world situation, which may operate on larger scales. Environmental impacts are limited and especially pertinent with a coastal habitat to the west and AONB to the east, to which the authors avoided growth. The game simulated a 'happy' community but only with distanced industrial zones and peripheral urban infrastructure facilities. Some existing real-world housing remained underutilised and abandoned in the game, such as a student housing block nestled against the West Coast mainline station, and heavy traffic and congestion were featured on the main arterial route to campus and the M6 motorway link through Galgate. Regional planning considerations are limited due to the game mechanics, but they could be an important area of investigation to stimulate interurban connectivity, growth, and tourism on a larger scale as well as incorporate a broader range of actors (Harrison et al., 2022).

Through gameplay, interesting scenarios could be visualised, such as connecting the canal system and River Lune and a new train station for new mobilities outside of the current real-world system and policy considerations. The game is more suited to local development plans than regional ones at present. Finally, once players are satisfied with modelled scenarios, a modification allows the export of the developed model as an OSM map via CimToGrapher for use in GIS software. Thus, the 'black box' of the game system provided ludic experiences, but deviations to the development policies regularly occurred. The possibilities of new urban forms and transport options outside the plan, such as a new train station, helped develop unconsidered options and delivered sustainable growth in the game.

Limitations

Henri Haimakainen, the game's designer for Colossal Order, Paradox Interactive, discusses the system of the sequel *Cities: Skylines II* and states that the game has five levels of educational attainment featured and mapped to school building types. Education is linked to job progress and happiness (Cities: Skylines, 2023). The simplified view of citizen behaviour indicates the reductive aspects of many real-world planning aspects due to the goal of playability. However, the game system management and UI of both game releases are highly intuitive. For example, a 'chirper' bot resembling social media narrates game progress, and there are expandable and collapsable dashboard indicators, utilising easily readable colour ramps and real-time situations, applied policies, and governance dashboards for complete overviews for players. This UI and the game mechanics

Figure 5.17
Gaming and the Future City, Campus in the City, Lancaster University, 2024. Photographer Robin Zahler.

provide a critical pedagogical tool in the game's planning and development system, with clear targets for progress, which is nonexistent in the range of UDT cases (Figure 5.17).

Often, the citizen-focused 3D experiences of UDTs, dashboards, and interaction do not help viewers navigate and understand the range of proposals and scenarios being presented or reference existing planning documentation, which normally consists of heavy text documents and PDFs, which is often assumed a priori knowledge. Thus, UDTs, even though adopting game technology, are one-way communication devices.

The UDTs do not explain *the system of planning* itself, unlike *Cities: Skylines* which provides data stories of the urban spaces being designed. Many of these planning-led UDTs could reflect on *Cities: Skylines*, which is transparent in terms of the *system of the game* and contains a rich range of explanations embedded in the game mechanics through the UI and bots. The game could be used to present a range of scenarios for the public to sample preferable options and schemes through future co-designed workshops such as naming new districts, blocks, and zones. Interest in the CBG genre should not be an immediate barrier for usage in workshops, given the game has an estimated 10.45m owners since its release, 69 extension packs, with around 57,000 peak users at a time (SteamDB, 2024). This future work also has the potential for replicability, given the computing needs for the game are not intense, given the game's release in 2015. Additional British modifications and housing typologies could also be utilised to increase degrees of realism and suitability in this case (Biffa, 2022).

The applied case was play-tested by three PhD Candidates at Lancaster University, Yuxuan Zhao, Lisha He, and Shuning Feng, in the development stages. In author interviews:

Shuning reflected,

- **I think about real-world mapping issues, such as connectivity and convenience of road layouts, and strive to circumvent these problems within my simulated cityscape; road planning is a complex element of the gameplay. Creating a city with efficient traffic flow, logical vehicle routing, and coherent connections between different zones is crucial to the game's appeal.**

Yuxuan reflected,

- **Using games to build realistic cities helps me think more clearly about area planning from a citizen's perspective, such as the location of train stations and shops around homes… it makes me more interested in urban life, and I feel the joy of urban design in it.**

And Lisha finally reflected,

- **This game is modelled after Western urban planning, where residences typically line the streets, contrasting with Chinese cities where residential areas often form enclosed complexes. Therefore, when playing this game, I adapted by planning the overall city road network and considering that residential zones can be different from the tracks in China.**
- **This game has taught me a lot about data visualisation and analysis. For example, it shows how vital traffic management is for a city's smooth operation and how public services and housing prices increase resident satisfaction. This analysis could be a practical way for teenagers to learn about urban planning, Geographic Information Systems (GIS), and Artificial Intelligence (AI). They can see how their designs turn into data analysis results and understand the complexities and diversity of urban planning.**

The applied *Cities: Skylines* case was play-tested with 60 children aged 7 to 13 in 'Gaming and the Future City' for two hours on university-supplied laptops as part of the annual campus in the City event (2024) co-delivered by a representative from Lancaster Council. The process was then repeated by playing Windermere, Lake District National Park, as part of the Windermere Science Festival with 60 children aged 7 to 13 in drop-in sessions (2024). Twenty adults also played at these events, totalling n.140, for a total of 280 hours of gameplay. The full results of this extensive playtesting cannot be reported in this chapter; however, post-event, to complete the workflow, all models created by participants have undergone GIS semantic classification of game images, which is the classification percentages of various elements in order to understand the perception and choice decisions and urban design. Post-event analysis has resulted in reports feedback to the local planning teams. This case has generated data as a result of playtesting in terms of how users respond and visualise their own future scenarios in relation to existing planning policies, creating highly participatory experiences that are currently lacking in the UK and elsewhere.

There are limitations to this method, generally, game players struggled to establish the infrastructure, water, roads, and power in the first game phase but soon progressed to various creative layouts and communities and recognised the geography of the area. In terms of adult gameplay, demographic engagement was mixed, with adults (parents who played at the event) enjoying the game with a prior interest in the CBG genre, such as *SimCity* and other adults struggling to engage with the game interface but received assistance from the event team and their children! Further gaming events with pre-designed real-world planning scenarios, including evaluation phases, are required to expand this research.

CONCLUSION

The emergent situation of UDTs and the use of games technologies for participatory planning has crossovers with digital games in terms of interfaces, 3D environments, and data visualisation. This chapter has argued that we need to better understand how particular design decisions change the nature and focus of the experience. Digital games provide data on how we play; in compassion, we also need data on the nature of interaction with UDTs. These interactions are fundamental and at the heart of social-technical relations. Digital games cannot replicate real-world planning systems; there is always an abstraction; however, with low public engagement in planning, there is a need to consider how game design approaches could play an essential role in increasing public interaction within current planning processes.

In the case of *Cities: Skylines,* the game experience contains embedded learning of the game system and its mechanics, whereas the sample of existing UDT cases in the urban game continuum is a one-directional information flow. The use of *Cities: Skylines* puts participants at the heart of the problem space. The participants generated data from play, which, in turn, loops back in to inform future scenarios. Creating playable experiences with embedded learning and progressive UIs could provide many higher-level modes of citizen engagement, reflections, and redesigns of the real-world planning system via serious games. There are limitations to this approach in that players are still utilising a range of pre-established game rules and dynamics, and there is a lack of opportunities to design an overall system in tandem with designers. In comparison, Damjan Jovanovic's (2021) *Planet Garden* is an open exploration terraform simulation game exploring non-linear systems in which players build and balance various sources for 'worldmaking'. The game tests underlying urban assumptions and allows players to design the system from the inside out for various future scenarios (Figure 5.18).

While players cannot explore the 'black box' of *Cities: Skylines* as a system, unlike *Planet Garden*, the extensive modification possibilities are useful, as is the pedagogic benefits, including players reflecting and thinking about networks, connections, and relationships with each micro city building planning decision played, especially for sustainable transit networks. This is a bottom-up approach, achieved in much more substantial ways in the AI-Powered Digital Twin for Sarajevo Urban Plan (Figure 5.19). The reflexive space via gaming is an incredibly useful device to promote discourse around future environments through the use of UDT systems rather than a one-way information source communicated via gaming technologies and interfaces.

Providing opportunities and games to explore 'problem landscapes' via UDTs can help the public scaffold their world-building, which is essential for planning engagement and participation practice. In a period in which cyber-physical sensors and models that constitute urban digital twins are increasingly being sought to virtualise our urban environments, we must introduce the ludic and not just the games technology to simulate future space and life.

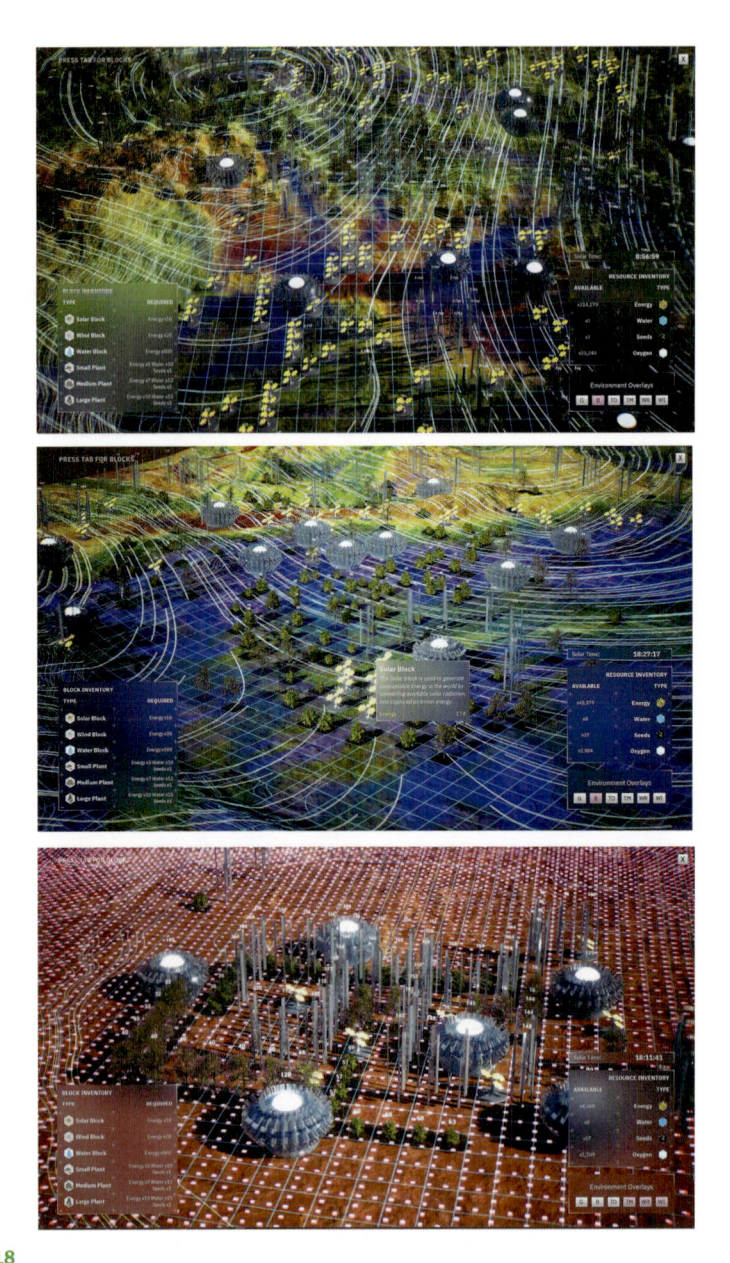

Figure 5.18
Damian Jovanovic, 2021, Planet Garden. Models: Nikhil Bang, Xiaolei Liu, Hakcheol Seo, Sanghyun Suh. Landscape Design: Ilaria Lu, Yaqing Mao, Caleb Roberts, Runhuan Wang. Vegetation. Mariam Aramyan, Wen Chen, Shuo Feng, Yehong Mi. Interface Design: Yuhong Gong, Robert Leising, Pan Tan, Game Manual: Eva Besmerti, Carlo Sturken. https://worldmaking.xyz/Planet+Garden/Planet+Garden.

CASE STUDY – AI-POWERED DIGITAL TWIN FOR SARAJEVO URBAN PLAN

To support and enable the urban transformation of the City of Sarajevo through quantitative and data-driven assessments, we developed the first AI-powered Digital Twin of the Canton of Sarajevo. Policymakers in Sarajevo are now using this novel methodology based on large-scale, agent-based, and bottom-up simulations to inform the design of the new urban plan.

As a multi-stakeholder project, the Urban Transformation Project Sarajevo (UTPS) is developed between the ETH Zurich Prof. Klumpner's Chair of Architecture and Urban Design, ETH Zurich Prof. Abhari's Laboratory of Energy Conversion, ETH Zurich spin-off SwissAI, University of Sarajevo and Canton Sarajevo Institute of Planning and Development. The overarching key component of the project is the elaboration of the urban plan for Sarajevo for the next 20 years.

DATA-DRIVEN AND AGENT-BASED SIMULATIONS

The UTPS project re-invents the application of the ETH simulation software EnerPol to support decision-making with data-driven and agent-based models. The development of a geospatial database, exhibiting unprecedented granularity, and calibrated with on-field measurements using the UTPS mobile laboratory «Studio Mobil» (e.g. wind measurements or traffic counts) ensures that EnerPol's data-driven assessments are accurate and comprehensive, covering all scales of the urban plan process.

Agent-based-modelling is a powerful application of AI algorithms in Digital Urban Planning. In EnerPol, agents representing people, buildings, or vehicles are characterised by a set of known input parameters and reinforced by machine learning predictive models. AI-powered agents adapt their behaviour by accounting for

GEODESIGN, URBAN DIGITAL TWINS AND FUTURES

Figure 5.19
Michael Walczak, Marco Pagani, 2024, Digital Twin's agent-based simulations visualised on a physical 3D model for a more immersive and participatory experience in the "Urban Design Studio Sarajevo" Project Space © Dr. Michael Walczak, Dr. Marco Pagani, ETHZ Prof. Klumpner Chair of Architecture and Urban Design.

252

Figure 5.20
The AI-Powered Digital Twin simulates country-wide public and private mobility at time resolution of one second ©
Dr. Michael Walczak, Dr. Marco Pagani, ETHZ Prof. Klumpner Chair of Architecture and Urban Design.

interactions with each other and the system, by learning from experience, and by optimising their choices over time. For example, in traffic simulations, AI enables drivers to learn optimal driving strategies to reduce traffic congestion.

In the Sarajevo case, we applied agent-based simulations in the assessments of infrastructure, private and public mobility, population, energy, decarbonisation, and urban developments.

FACILITATING DECISION-MAKING

It is challenging to understand the impact of urban planning on a city's functioning. Various dynamics, such as infrastructure usage, are nowadays deeply interconnected. The holistic approach of Digital

Figure 5.21
Public Hearing on the new urban plan at the Sarajevo Olympic Center supported by results of the Digital Twin (projected)
© Photo: Amina Ahmetagić, ETHZ Prof. Klumpner Chair of Architecture and Urban Design.

Twins is best suited to decrypt complexity. Using 4D real-time visualisation and game-engine techniques, urban planners can assess the impact of future scenarios on several externalities related to urban comfort, such as pollution, noise, traffic congestion, accessibility to public services, and ecosystem preservation.

In the UTPS project, integrating ABM with AI provided therefore valuable insights into the long-term consequences of urban planning decisions. The Digital Twin was adopted by the Cantonal Planning Office Sarajevo as the new platform to engage with citizens and local authorities in public hearing formats (Figure 5.21). The predictive algorithms are applied by local stakeholders to forecast, evaluate, and compare the key performance indicators of future urban planning scenarios, effectively raising the threshold of digital literacy among local stakeholders.

The ultimate goal of the presented methodology is to support decision-making processes for evidence-based policymaking, in particular for the spatial plan, urban plan, regulatory plans, and small-scale interventions in Sarajevo.

CHAPTER 5
GAMING, WORLD-BUILDING, AND PARTICIPATORY PLANNING

Figure 5.22
Real-time game engine visualisation of new City Center development in Sarajevo Rajlovac. © Dr. Michael Walczak, ETHZ Prof. Klumpner Chair of Architecture and Urban Design.

- Dr. Michael Walczak, Postdoctoral Researcher, ETH Zurich Klumpner Chair of Architecture and Urban Design
- Dr. Marco Pagani, Postdoctoral Researcher, ETH Zurich Klumpner Chair of Architecture and Urban Design.

Digital Twin Sarajevo by the Urban Transformation Project Sarajevo (UTPS) © Film: Dr. Michael Walczak, ETHZ Prof. Klumpner Chair of Architecture and Urban Design. https://youtu.be/r84BamC46fo?feature=shared

REFERENCES

5PM Studio. (2021). *The Architect: Paris* [Digital game]. Endo Games.

Abt, C.C. (1970). *Serious games*. University Press America.

Addis, M. (1964). *The Summerian Game* (IBM 7090) [Digital game]. BOCES; IBM.

Adopted Policies Map – July 2020. (2020a). Lancaster City Council. Retrieved May 21, 2024, from https://lancaster.maps.arcgis.com/apps/webappviewer/index.html?id=8a956391c7ee4c68a74b31f3732476cb.

Bailrigg *Garden Village, Lancaster*. (2022). JTP. Retrieved March 5, 2022, from https://www.jtp.co.uk/projects/bailrigg.

Blank, G. (1978). *Santa Paravia en Fiumaccio* [Digital game]. Instant Software.

Bereitschaft, B. (2016). Gods of the city? Reflecting on city building games as an early introduction to urban systems. *Journal of Geography*, 115(2), 51–60. https://doi.org/10.1080/00221341.2015.1070366.

Biffa. (2022, January 6). *Biffa's British city build – Assets (2022)*. Steam Community. https://steamcommunity.com/sharedfiles/filedetails/?id=2710090707.

Bogost, I. (2011). *How to do things with videogames*. University of Minnesota Press.

Bradfield, R., Wright, G., Burt, G., Cairns, G., & van der Heijden, K. (2005). The origins and evolution of scenario techniques in long range business planning. *Futures*, 37(8), 795–812. https://doi.org/10.1016/j.futures.2005.01.003.

Butler, R., Garnet, L., Shah, P., & Krabbe, I. (2020). *The future of engagement*. RTPI. https://www.rtpi.org.uk/media/7258/the-future-of-engagement.pdf.

Caseous Stranger. (2022, February 14). *Pittsburgh 1:1 mod list*. Steam Community. https://steamcommunity.com/sharedfiles/filedetails/?id=2753745134.

Celestial Software. (1991). *Utopia: The Creation of a Nation* [Digital game]. Gremlin Interactive.

City Planner Place. (2022, April 1). *Recreating the first garden city using the real city plans in Cities Skylines | Timelapse | UK Build*. Steam Community. https://steamcommunity.com/sharedfiles/filedetails/?id=2787603729.

Cities: Skylines. (2023, August 31). *Simulating life | Developer insights Ep 11 | Cities: Skylines II* [Video]. YouTube. https://youtu.be/OTOPkhWcIRQ?feature=shared.

Colossal Order. (2015). *Cities: Skylines* [Digital game]. Paradox Interactive.

Colossal Order. (2023). *Cities: Skylines II* [Digital game]. Paradox Interactive.

Crooks, A., Heppenstall, A., Malleson, N., & Manley, E. (2021). Agent-based modeling and the city: A gallery of applications. In W. Shi, M.F. Goodchild, M. Batty, M.-P. Kwan, & A. Zhang (Eds.), *Urban informatics* (pp. 885–910). Springer.

Cureton, P & Coulton, P 2024, Game-Based Worldbuilding: Planning, Models, Simulations and Digital Twins, *Acta Ludologica*, vol. 7(1). 10.34135/actaludologica.2024-7-1.18-36

Daglow, D. (1982). *Utopia* [Digital game]. Intellivision; Mattel Aquarius.

Daglow, D.L. (2018). *Indie games: From dream to delivery*. Sausalito.

Daye, C. (2019). *Experts, social scientists, and techniques of prognosis in Cold War America*. Springer Nature.

Devisch, O. (2008). Should planners start playing computer games? Arguments from SimCity and Second Life. *Planning Theory & Practice*, 9(2), 209–226. https://doi.org/10.1080/14649350802042231.

Djaouti, D.A. (2011). Origins of serious games. In M. Ma, A. Oikonomou, & L.C. Jain (Eds.), *Serious games and edutainment applications* (pp. 25–43). Springer.

Dyment, D. (1968). *Hamurabi* (FOCAL) [Digital game]. Digital Equipment Corporation.

Energia- ja ilmastoatlas [Energy and Climate Atlas]. (n.d.). Helsinki Region Infoshare. Retrieved May 21, 2024, from https://kartta.hel.fi/3d/atlas/#/.

Focus Entertainment. (2013). *Cities XL* [Digital game]. Monte Cristo; Pullup Entertainment.

Gaber, J. (2007). Simulating planning: SimCity as a pedagogical tool. *Journal of Planning Education and Research*, 27(2), 113–121. https://doi.org/10.1177/0739456X07305791.

Harrison, J., Galland, D., & Tewdwr-Jones, M. (2022). *Planning regional futures*. Routledge.

Hunicke, R., Leblanc, M., & Zubek, R. (2004). MDA: A formal approach to game design and game research. In D. Fu, S. Henke, & J. Orkin (Eds.), *Proceedings of the AAAI workshop on challenges in game AI* (pp. 1–5). AAAI. https://cdn.aaai.org/Workshops/2004/WS-04-04/WS04-04-001.pdf.

Imam, A., & Lahoud, C. (2021). *The block by block playbook: Using Minecraft as a participatory design tool in urban design and governance*. United Nations Human Settlements Programme (UN-Habitat). https://unhabitat.org/sites/default/files/2021/09/1-bbb_playbook_publication_final.pdf.

Jolly, R., & Budke, A. (2023). Assessing the extent to which players can build sustainable cities in the digital city-builder game "Cities: Skylines". *Sustainability*, 15(14). https://doi.org/10.3390/su151410780.

Jovanovic, D. (2021). *Planet Garden* [Digital game]. https://worldmaking.xyz/Planet+Garden/Planet+Garden.

Juraschek, M., Herrmann, C., & Thiede, S. (2017). Utilizing gaming technology for simulation of urban production. *Procedia CIRP*, 61, 469–474. https://doi.org/10.1016/j.procir.2016.11.224.

Kahn, H. (1967). The next thirty-three years: A framework for speculation. *Daedalus*, 96(3), 705–732. https://www.jstor.org/stable/20027066.

Khan, T.A., & Zhao, X. (2021). Perceptions of students for a gamification approach: Cities Skylines as a pedagogical tool in urban planning education. In D. Dennehy, A. Griva, N. Pouloudi, Y.K. Dwivedi, I. Pappas, & M. Mäntymäki (Eds.), *Responsible AI and analytics for an ethical and inclusive digitized society* (pp. 763–773). Springer. https://doi.org/10.1007/978-3-030-85447-8_64.

Latour, B. (2004). Why has critique run out of steam? From matters of fact to matters of concern. Critical Inquiry, 30(2), 225–248.

Levelling-up and Regeneration Bill: Reforms to National Planning Policy. (2023, December 19). GOV.UK. https://www.gov.uk/government/consultations/levelling-up-and-regeneration-bill-reforms-to-national-planning-policy/levelling-up-and-regeneration-bill-reforms-to-national-planning-policy#chapter-10--national-development-management-policies.

Local Plan Part One: Strategic Policies & Land Allocations DPD. (2020b, July 29). Lancaster City Council. Retrieved February 14, 2021, from https://www.lancaster.gov.uk/planning/planning-policy/land-allocations-dpd.

Manns, S. (2017, May 10). *Planning and public engagement: the truth and the challenge*. RTPI. https://www.rtpi.org.uk/blog/2017/may/planning-and-public-engagement-the-truth-and-the-challenge/.

Mapping Morrisville: Land Use Plan. (2021, February 23). Houseal Lavigne Associates. Retrieved May 21, 2024, from https://www.hlplanning.com/portals/morrisville/.

Maxis. (1989). *SimCity* [Digital game]. Maxis.

Maxis. (1992). *SimRefinery* [Digital game]. Maxis.

Maxis. (1993). *SimCity 2000* [Digital game]. Maxis.

Milgram, P., Takemura, H., Utsumi, A., & Kishino, F. (1995) Augmented reality: A class of displays on the reality-virtuality continuum. In H. Das (Ed.), *Proceedings volume 2351: Telemanipulator and telepresence technologies* (pp. 282–292). SPIE. https://doi.org/10.1117/12.197321.

Nass, S., & Weeks, O. (2022). *SimNimby* [Digital game]. Itch.io. https://opoulos.itch.io/simnimby.

Olszewski, R., Cegielka, M., Szczepankowska, U., & Wesolowski, J. (2020). Developing a serious game that supports the resolution of social and ecological problems in the toolset environment of Cities: Skylines.

ISPRS International Journal of Geo-Information, 9(2), 1–20. https://doi.org/10.3390/ijgi9020118.

Pearce, C. (2022). Sims, BattleBots, Cellular Automata God and Go: A conversation with Will Wright by Celia Pearce. *Game Studies*, 2(1). https://www.gamestudies.org/0102/pearce/.

Mojang Studios. (2011). *Minecraft* [Digital games]. Mojang Studios.

Plethora Project. (2016). *Block'hood* [Digital game]. Devolver Digital.

Poplin, A., de Andrade, B., & de Sena, Í. (2020). Geogames for change: Cocreating the future of cities with games. In D. Leorke, & M. Owens (Eds.), *Games and play in the creative, smart and ecological city* (pp. 64–94). Routledge. https://doi.org/10.4324/9781003007760.

Purple Yonder. (2022). *Little Cities* [Digital game]. nDreams.

Salvador, P. (2020, May 19). *When SimCity got serious: The story of Maxis Business Simulations and SimRefinery*. The Obscuritory. https://obscuritory.com/sim/when-simcity-got-serious/.

Starr, P. (1994). Seductions of sim: Policy as a simulation game. *The American Prospect*, 1994(17), 19–29.

SteamDB. (2024). *Cities: Skylines*. SteamDB. Retrieved February 10, 2024, from https://steamdb.info/app/255710/charts/#all.

System 3. (1997). *Constructor* [Digital game]. Acclaim Entertainment.

Terzano, K., & Morckel, V. (2017). SimCity in the community planning classroom: Effects on student knowledge, interests, and perceptions of the discipline of planning. *Journal of Planning Education and Research*, 37(1), 95–105. https://doi.org/10.1177/0739456X16628959.

Thinking Tools. (1994). *SimHealth* [Digital Game]. Maxis.

Wicaksana, G.B.A., & Darmawan, G.S. (2022). Geographic simulation of Denpasar city in game cities: Skylines. In I.M. Suwitra, I.N. Nurjaya, P.D. Astuti, A.D.Y. Pratama, & R. Rahim (Eds.), *Proceedings of the 1st Warmadewa international conference on science, technology and humanity, WICSTH 2021* (pp. 3–8). https://doi.org/10.4108/eai.7-9-2021.2317763.

Yeh, A.G.O., Li, X., & Xia, C. (2021). Cellular automata modeling for urban and regional planning. In W. Shi, M.F. Goodchild, M. Batty, M.-P. Kwan, & A. Zhang (Eds.), *Urban informatics* (pp. 865–883). Springer.

Young, G.W., & Kitchin, R. (2020). Creating design guidelines for building city dashboards from a user's perspectives. *International Journal of Human-Computer Studies*, 140. https://doi.org/10.1016/j.ijhcs.2020.102429.

CONCLUSION — GEODESIGN, URBAN DIGITAL TWINS, AND FUTURES AT THE EDGE

CONCLUSION

This book has charted the emergent possibilities of urban digital twins (UDTs), which the author defines as digital representations at a set fidelity of physical element(s), including its behaviour, which is connected and integrated for efficiency. UDTs cover three broad areas: people, models, and platforms. These broad themes were used to design an overarching frame for the book. UDTs are the next iteration of smart city developments and can move beyond the criticisms and failures of 'smartness.' UDTs were utilised as a term to escape global city-centric views and to democratise and cover a wide variety of urban typographies and scales. The use of the term UDT is particularly important; it is a term for inclusiveness in this new emergent systems approach. UDTs are an embryonic area that offers an assembly of sensors, models, near-real-time data, and analytical capabilities that are organised and designed for various urban applications, including energy, transport, and planning. Yet, there are fundamental questions on the direction of UDTs, with many discussions around UDT 'optimisation' but less so covering the climate emergency. What territories should UDTs cover? What data should be assembled? How are these systems structured, and how are stakeholders communities and decision-making derived? How should UDTs be organised for sustainability and related to citizen engagement?

The book's aim has been to address these questions or further reduce the gaps in knowledge. Practically, each chapter is supported by tutorials that address varying aspects of UDTs on the companion website, which will be updated regularly. The developmental aspects of UDTs are areas of concern. To date, they have concentrated on global cities, purporting to offer a range of efficiencies in planning and mobilities amongst other sectors (WEF, 2022). However, there has been limited consideration of the technological aspects of UDTs for other spatially significant regions and territories. This raises the question of whether UDTs applied to global cities are a misdirection of application and whether they could result in a repeat of the earlier failures of first-generation ICT-centric smart cities (Datta, 2015; Kitchin, 2018). Moreover, Chapter 1 outlined the climate emergency and global urbanisation, which, for complex geographical areas, does not lend itself to straightforward techno-future solution-based approaches. At this point, in terms of new developments in urban science, we need to design UDTs with a core purpose for meaningful change and climate adaptation.

Geodesign is an open framework and process developed from early experiences and the modern development of GIS. As Shannon McElvaney states, "the goal that a geodesign framework infuse design with a blend of value and science-based information made relevant by its geography and history to help designers and stakeholders make the wisest decision possible, taking into account potential impacts" (McElvaney, 2012, p. 12). This book has argued that we do not need novel approaches but combine emergent systems from UDTs with a collaborative approach that has been around for over 30 years and provides an extensive evidence base for decision-making for spatial change. The novelty, in this case, arises from the combination of socio-technological factors. The current state-of-the-art UDTs utilise 3D GIS and common city information models (CIMs) or collated experiments exploring individual components of a UDT system. The range of works has real potential, though there is an important organisational aspect of how these experiments are conducted. Often, state-led national agencies and data providers commission work via a top-down approach. However, this does not necessarily answer one of the main UDT challenges, which is the frequency of data collection, which can offer 'near real-time' results. This organisational approach also raises questions about who they are UDTs for. This book suggests that citizen science and large-scale volunteered geographic information (VGI) data can only address this research challenge as a bottom-up approach using the cheapest, most ubiquitous sensor on the planet, the smartphone. VGI is not without challenges due to data quality, but VGI could be fostered if the purpose of UDTs was much clearer and tangible benefits were communicated. Design fiction, as a speculative design research method, was offered to provide foresight and futures via a world-building approach to map alternatives through diegetic prototypes and scaffolding. In a period of extensive artificial intelligence (AI) experimentation, design fiction is particularly important in order to make the ethical dimensions of AI in the built environment but also upskilling of professionals and democratisation of AI. There are large-scale policy challenges in emergent UDTs which still require understanding. Through a case study of electrifying transport in the Cambridge, region, UK, Timea Nochta, Li Wan, Jennifer Schooling, and Ajith Parlikad rightly argue that,

> **Characteristics of urban governance, including the use of model outputs as evidence in policy decision-making must be considered [...] Moreover, modelers (and the supply side more generally) must engage with a diverse set of societal actors, as well as a wider spectrum of policy alternatives and modeling approaches, in order to be able to provide contextually relevant and appropriate insights and recommendations**
> (Nochta et al., 2020, pp. 282–283)

Such a reciprocal relationship requires interdisciplinary working and collaboration which the geodesign framework can potentially foster. CIMs constitute the current paradigm and are precursors to UDTs. They have a number of similarities and should be focused and targeted for specific geographic needs and policy goals. However, they often do not yet have two-way communications or connected sensing and predictive capabilities, as we have seen in the various maturity models (Chapter 2). However, CIMs and UDTs hosted on cloud platforms are being integrated around the world as part of digital transformation in planning, which offers volumetric mapping of environments, which provides an evaluation space on the critical 3D dimensions of the built environment in order that macro and micro decisions can be formulated. The achievements in UDTs revolve around a number of interrelated sub-components and schematics, which have been defined as baseline data, data dashboards, and interactive elements such as extended reality (XR). The Virtual Helsinki case is particularly important organisationally, acting as an umbrella to explore many of the aspects of a UDT, such as CIMs in construction, strategic energy planning for climate change, and citizen engagement in major planning development.

Given that the turn of the 20th century involved the first civilian aerial surveys and a significant change of cartographic mapping based on land surveys to aerial photography, innovation has rapidly continued leading to today's cartographic practice, in which aerial data can appear at high frequency via a cloud platform, such as the case of Aerometrix (Figure 6.1). The progress of remote sensing is a significant advancement. Aerometrix provides a range of datasets from

Figure 6.1
Three-dimensional Photogrammetry of Adelaide Oval, South Australia by Aerometrex, rendered in Unreal Engine, 2024.

aerial acquisition, in this case, brought into a 3D baseline for a UDT. Indeed, Michael Batty has carefully mapped many of these developments and the advancement of urban science and computation as techniques and technologies diffuse and coalesce in high-frequency cities (Batty, 2024, Chapter 16). Mental challenges remain in the organisation and collection of data for UDTs in terms of the cost of collection, access, and use of GeoAI: AI techniques such as machine learning or generative AI. This book has avoided the particular implications of AI due to the focus on UDTs and the significant scope of the topic. However, benefits are likely widespread in various modelling approaches and choices for UDTs, such as procedural methods as found in Houseal Lavigne, Jardin Siti project (Figure 6.2). ABM, or Michael Batty and Richard Milton's QUANT model for large-scale 'what if' scenarios (Batty & Milton, 2021). Suppose UDTs are assumed to be federated systems. In that case, they will contain the best models we have; yet to an extent, they may

Figure 6.2
Houseal Lavigne, Jardin Siti (Garden City), 2023. Revising Ebenezer Howard's Garden City, a comprehensive graph network was created, and procedural buildings and elements were mapped to different districts, transit networks were established, and land uses defined. Jardin Siti was constructed with Esri's ArcGIS CityEngine, Nvidia's Omniverse, and Epic's Unreal Engine. Jardin Citi Story Map: https://storymaps.arcgis.com/stories/0d588edf9c19 4cd7a378f55d 77c3bd86.

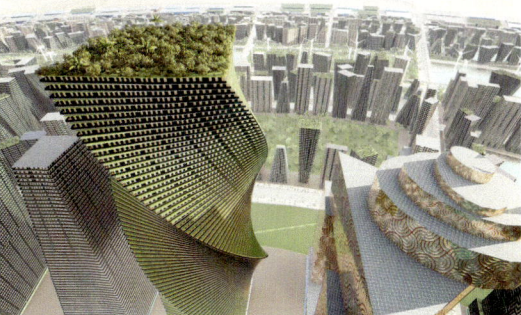

be abstracted beyond real use or lack the validation required to be 'trustworthy,' which is a particular issue in socio-technical relations for UDTs. Moreover, the specialisation and modelling approaches need to be democratised, particularly in urban planning contexts, in which the upskilling of workforces will become critical; this challenge was faced by the author's Lancaster City Information Model (LCIM) pilot in training planners to use the model and data created in a prototype. In seeking replicas of the real world through a UDT systems approach, laypersons must fundamentally *believe* them.

To foster a usable and applicable UDT, geodesign is critical in bringing together a range of stakeholders in order to envision and experiment with data. Geodesign, as a framework, is delivered as six procedural iterations: representation model, process model, evaluation model, change model, impact model, and decision model, and Chapter 3 contributes to knowledge through an open Card UX pack in English and Chinese versions along with game boards that were tested at the CIM Forum. The application of geodesign is not an isolated case, and many of the applications of geodesign across the world are charted in Chapter 3. Eleven applicable methods were established alongside mapped cases to further establish the relationship between UDTs and geodesign, covering people, models, and platforms (Chapter 4). There are no prescriptive methods for geodesign and this is one appeal of the framework, allowing flexibility and adoption across geographies. For example, Flora Roumpani's use of CityEngine procedural modelling of four scenarios is particularly pertinent in the change model stages of geodesign (Figure 6.3).

One component of UDTs, as constituting people, models, and platforms, is the use of game engines for simulation and the visualisation of urban environments. Chapter 5 discussed the history of city-building genres in digital games and their relationship with real-world planning. In the Kriegsspiel war gaming example Figure 6.4, at the Innovation Laboratory at the Bundeswehr Command and Staff College, Hamburg Germany, war gaming is very much a part of

Figure 6.3
Flora Roumpani, 2022. "Scenarios of urban growth using a gravity type model in CityEngine. From left to right, predictions of organic growth for 2016, 2017, and 2018" (2022, p. 327).

leadership development and focused upon critical thinking. Digital games are abstracted simulations of complex real-world spaces, yet the design decisions we make are critically important for urban planning, just as in wargaiming. Games are highly playable, enjoyable, data generating and participatory. The contribution to knowledge here is that UDTs utilising game technology could significantly enhance their participatory aspect and citizen engagement using game design principles and processes such as flow, game mechanics, progression, shell, and in-game screens (equivalent to data dashboards). For example, often, there is limited understanding of sub-surface assets and the geology of urban areas. The British Geological Survey sought to overcome this problem through play by hosting a national dataset in the digital game Minecraft (Figure 6.5).

GEODESIGN, URBAN DIGITAL TWINS AND FUTURES

Figure 6.4
Kriegsspiel Game hosted by Professor Jorit Wintjes of the Conflict Simulation Group at the University of Würzburg, Lieutenant Colonel (GS) Thorsten Kodalle, Innovation Laboratory at the Bundeswehr Command and Staff College, Hamburg Germany.

Figure 6.5
British Geological Survey (BGS), To develop engagement with sub-surface mapping, the BGS released a geological model in the digital game Minecraft.

Such communication and dissemination can help highlight often under-considered aspects of the built environment and form part of the efforts to map underground utilities at scale across the United Kingdom through the National Underground Asset Register (NUAR). Many other countries are conducting important geospatial experiments in this vein (DSIT, 2022).

Chapter 5 argued for a move beyond the limited use of games technology and the inclusion of a games design approach through a taxonomy of city-building games and real-world UDTs, mapping their attributes based on four values. An applied case using a city-building game and imported real-world geodata with children and adults (n.140) and planners in a workshop was conducted in order to create alternative visions in a digital game for development on a real-world site allocated for development in the UK. This method could be replicated at a low cost in other areas with data shortages and limited resources, which is a key feature of the Block-by-Block playbook and method (Chapter 4). The use of virtual and interactive experiences is an important aspect of urban digital twins. Just as UDTs must have two-way data interactions between connected sensors and the system, so should this two-way interaction apply to publics. In the Mid Cornwall Metro project in the UK, a VR walkthrough was created by Digital Urban in order that the audience could take a balloon ride of the scheme (Figure 6.6). In another example, Alenka Poplin et al. have utilised geodesign process and geogames for youth engagement that reflects a real-world planning issue and developed a number of prototypes, including *Tirolcraft*, in Santa Leopoldina, Espírito Santo, Brazil concentrating on heritage values and architectural materials as well as landscape typologies (Poplin, 2022) (Figure 6.7).

The act of world-building using games is particularly important as it can reflect players' own values, as well as explore underlying philosophies. In Walden, a game, the naturalist Henry David Thoreau's philosophy is presented through an open-world simulator in which players experience landscape change of seasons (Figure 6.8). UDTs cannot be neutrally presented systems; they are reflectors of the

Figure 6.6
Digital Urban, Mid-Cornwall Metro, Mid-Cornwall Metro public engagement event, Newquay Cornwall, February 2024. Cornwall Council, GWR and Network Rail. Photographer Tom Last.

Figure 6.7
Alenka Poplin, Bruno de Andrade, Ítalo de Sena, 2022. Minecraft simulation game Tirolcraft, a serious game for 4- to 11-year-olds using geodesign cards, to record heritage values, Tirol district in the municipality of Santa Leopoldina, Espírito Santo, Brazil.

Figure 6.8
Walden, a game, Tracy Fullerton, and the Walden Team (USC Game Innovation Lab), 2023. The digital game translates the experience of naturalist Henry David Thoreau's stay at Walden Pond in 1845–1847. https://www.waldengame.com/.

political mechanism and policies in which they can be enabled; they are fundamentally entwined; like Walden, a game, there should be consideration of the basis of formation and consideration of a change of philosophy and direction. In a planning context, as Alexander Wilson and Mark Tewdwr-Jones state,

> *Being prepared to listen to, understand, and engage with communities' experiential lived accounts of the built environment will be a challenge for planners who have been used to taking a narrow, planning as land use and development, perspective of their roles.*
> (Wilson & Tewdwr-Jones, 2021, p. 81)

UDTs are high-frequency systems reflecting the complexity of the urban environment; the participatory aspect of these systems should not be one-way; in fact, through digital games and the act of the play, data is generated, providing important feedback loops to a UDT system. UDTs can be used as systems for rigorously envisioning the futures we want and those we do not through deliberative geodesign. Critiques of 'city' digital twins have analysed the control room and surveillance-state micromanagement that could come to the fore in future systems (Charitonidou, 2022; Lehtola et al., 2022). However, it is the author's view that there is still an inclusive space for designing, experimenting, and playing scenarios through computation, UDTs, and real conversations with publics to turn towards the climate emergency.

Toolkits and Games	
Walden a game	https://www.waldengame.com/#home-section
UN Habitat	https://unhabitat.org/planning-for-climate-change-a-strategic-values-based-approach-for-urban-planners-toolkit
British Geological Survey, Minecraft Data	https://www.bgs.ac.uk/discovering-geology/maps-and-resources/maps/gb-geology-minecraft-world/

VISIONING AND EXPERIMENTATION

The Singapore Land Authority has established the most progressive digital twin to date, high on the TRL levels discussed in Chapter 1, and Virtual Singapore has been discussed at length (Therias & Rafiee, 2023). In the Urban Redevelopment Authority gallery space, the public can examine future plans and a range of 'smart' technologies, and there is a central area model. The physical model has unallocated zones for development in which children can build Lego models and place them on the main architectural model; sometimes, the designs are stronger than the global architecture companies. This is a small act, but it is one in which children participate to a small degree in a future vision (Figure 6.9). Visioning can take many forms, but UDTs reflect real-world spaces, they are the urban labs that provide a constellation of data, models, and platforms. The range of researchers involved in this space have a duty as Per Linde and Karin Book state,

Figure 6.9
Urban Redevelopment Authority (URA), Singapore Central Area Model. Alamy Stock. https://www.ura.gov.sg/Corporate/Singapore-City-Gallery.

> *To stage a design process in which we can, together with users, explore the new possibilities of integrating digital media services and content with the city landscape and shape place-centric interactions for city dwellers, commuters, and visitors.*
>
> (Linde & Book, 2014, p. 282)

While Linde and Book delivered and experimented with a new media project in Malmo, Sweden, their statement is particularly important in relation to UDTs. The form of interaction for an emergent field of urban science establishing UDTs will become critical. It is worth briefly describing three futures based on film media and relating this to visions for UDTs. This futuring aspect has been conducted with Nick Dunn in previous work (Dunn & Cureton, 2020).

First, Keiichi Matsuda's speculative design film Hyperreality (2016) showcases an augmented reality future in Medellín, Colombia, in which informatics and HCI create a society experiencing visual overload. The film provides core lessons for data visualisation and the representational challenges of representing multiple sensors and high-frequency data of urban spaces (Figure 6.10). UDTs are a system of systems that are carefully designed, transparent, and curated towards key policy decisions addressing the most pressing urban problems, not just standard routes.

Secondly, in another future city vision, Planet City, a film, VR experience, and book, explores a more dense city which has formed for ten billion people as part of the alleviation of climate disasters based on urbanisation statistics, the ideas of sprawl as found in the writing of William Gibson and total urbanisation idea of Constantinos Doxiadis and the concept of giving over of half the world to the wilderness by biologist Edward O Wilson (Young, 2021, pp. 33–36) (Figure 6.11). Liam Young's vision of managed retreat relies on highly developed technologies supporting the superstructure. It could show a UDT at its maximum TRL and maturity regulating air systems, agrivoltaics (dual solar energy and agriculture land use) and other core life systems.

Figure 6.10
Keiichi Matsuda, Hyper Reality, 2016. The speculative design by Keiichi Matsuda presents a kaleidoscopic future city in Medellín, Colombia in which physical and virtual realities have merged creating overloaded UI and interactions. http://hyper-reality.co/.

GEODESIGN, URBAN DIGITAL TWINS AND FUTURES

Figure 6.11
Liam Young, Planet City, 2021. Director Liam Young Production Design. Liam Young Costume Director/Producer Ane Crabtree VFX Supervisor Alexey Marfin VFX Case Miller, Aman Sheth, Vivian Komati, Yucong Wang Original ScoreForest Swords (Matthew Barnes) Vocals EMEL Lead Researcher Case Miller Researcher Pierce Myers Narrative Consultant Jennifer ChenWest Coast Costume Assistant Courtney Mitchell East Coast Costume Assistant Ela Erdogan Costume Artists Holly McQuillian, Karin Peterson, Kathryn Walters (Zero Waste Weavers), Aneesa Shami (High Altitude Bot Herder), Yeohlee Teng (Code Talker), Courtney Mitchell (Beekeeper), Ane Crabtree (Nomadic Worker, Algae Diver, Drone Shepard) Fibre Artist Janice Arnold Mask Artists Liam Young (High Altitude Bot Herder, Code Talker, Algae Diver, Drone Shepard), Zac Monday (Zero Waste Weavers), Aneesa Shami (Zero Waste Weavers) Costume Still Photography Driely S. Costume stills Photoshoot Produced by Eva Huang Performers David Freeland Jr, AJ and Miguel Alejandro Lopez, Joy Brown Commissioned by NGV Melbourne & Ewan McEoin.

CONCLUSION

Young's vision highlights the social issues to which a UDT could be deployed. To maintain its function as a cyber-physical system, there will be challenges in how UDTs adapt to new organisational approaches, changing disciplines, legislative structures, and political changes, and the viability and validity of UDTs will come under questioning.

Thirdly, in another speculative vision, the film Ready Player One, based on Ernest Cline novel, presents a mundane world in which the public escapes through haptic VR suites (Figure 6.12). The film is reductive and full of 1980s games' cultural references and nostalgia but raises questions about purely techno-mediated futures (Nordstrom, 2016). While there may be an increase in virtual worlds and metaverses, and this may increase escapism mediated via game technology, a UDT is fundamentally a mirror of the real world and entirely different. The degree of replication may come into question, but the replication

Figure 6.12
Ready Player One, 2018. American film co-produced and directed by Steven Spielberg, written by Zak Penn and Ernest Cline, and based on Cline's 2011 novel of the same name. The film stars Tye Sheridan, Olivia Cooke, Ben Mendelsohn, T.J. Miller, Simon Pegg, and Mark Rylance. Warner Bros/Alamy Stock.

277

must be of highly individualised urban settings which is counter productive for large scale homogeneous implementation. A UDT could be used to explore a number of speculative design fictions, but this would always resort back to informing the ability of the replica of the system itself. A UDT must have ludic experiences and predictive capability to help society map future directions, which is a significant challenge. These three examples presented three speculative tropes for a future built environment in which high-frequency interfaces are ubiquitous; urbanisation becomes hyper-dense supported by high-level integrated mature technologies and a warped techno-mediated future defining a purpose for interaction in virtual environments. Examining such media and speculations is useful in charting where we are heading via a backcasting and taxonomic approach. It also provides three examples of the many possible worlds in the future in which a UDT could potentially function.

We need to understand the forms of collaboration for people via geodesign for human-centric UDTs while at the same time embracing more-than-human perspectives (Yoo et al., 2023). UDTs should be deployed in highly individualistic ways, mirroring realities and rational complexities, not through blanket deployments, reflecting the geographies and parameters of their citizens. Urban modelling efforts should continue and be interlinked with multidisciplinary researchers. UDTs should be designed from the top down and the bottom up. The choices of 'twinning' should also be open and co-created. UDTs should be supported by federated models at high frequencies that can simulate and predict urban futures for climate mitigation. The range of UDTs should be designed through systems architecture and platforms that engage their citizens, generating feedback loops through acts of democratic play and engagement again through geodesign procedures. If the aforementioned actions can be achieved for UDTs facing the primary challenge of addressing socio-technical relations across people, models, and platforms, then, just then, UDTs will foster a paradigm shift in urban science to which, right now, we are at the edge.

REFERENCES

Batty, M., & Milton, R. (2021). A new framework for very large-scale urban modelling. *Urban Studies*, 58(15), 3071–3094. https://doi.org/10.1177/0042098020982252.

Batty, M. (2024). *The computable city histories, technologies, stories, predictions*. MIT Press.

Charitonidou M. (2022). Urban scale digital twins in data-driven society: Challenging digital universalism in urban planning decision-making. *International Journal of Architectural Computing*, 20(2), 238–253. https://doi.org/10.1177/14780771211070005.

Yoo, D., Bekker, T., Dalsgaard, P., Eriksson, E., Fougt, S.S., Frauenberger, C., Friedman, B., Giaccardi, E., Hansen, A.-M., Light, A., Nilsson, E.M., Wakkary, R., & Wiberg, M. (2023). More-than-human perspectives and values in human-computer interaction. In *Extended abstracts of the 2023 CHI conference on human factors in computing systems (CHI EA '23)* (pp. 1–3). Association for Computing Machinery, New York, NY, Article 516. https://doi.org/10.1145/3544549.3583174.

Datta, A. (2015). 100 smart cities, 100 utopias. *Dialogues in Human Geography*. https://doi.org/10.1177/2043820614565750.

Department for Science, Innovation and Technology and Geospatial Commission, National Underground Asset Register (NUAR). (2022, March 22). https://www.gov.uk/guidance/national-underground-asset-register-nuar.

Dunn, N., & Cureton, P. (2020). *Futures cities: A visual guide*. Bloomsbury Publishing.

Kitchin, R. (2018). Reframing, reimagining and remaking smart cities. In C. Coletta, L. Evans, L. Heaphy, & R. Kitchin (Eds.), *Creating smart cities* (pp. 219–230). Routledge.

Lehtola, V.V., Koeva, M., Elberink, S.O., Raposo, P., Virtanen, J., Vahdatikhaki, F., & Borsci, S. (2022). Digital twin of a city: Review of technology serving city needs. *International Journal of Applied Earth Observation and Geoinformation*, 114, 102915. https://doi.org/10.1016/j.jag.2022.102915.

Linde, P., & Book, K. (2014). Performing the city: Exploring the bandwidth of urban place-making through new-media tactics. In P. Ehn, E.M. Nilsson, & R. Topgaard (Eds.), *Making futures: Marginal notes on innovation, design, and democracy* (pp. 277–302). The MIT Press. https://www.jstor.org/stable/j.ctt9qfb58.18.

McElvaney, S. (2012). *Geodesign: Case studies in regional and urban planning*. Environmental Systems Research Institute.

Nochta, T., Wan, L., Schooling, J.M., & Parlikad, A.K. (2020). A sociotechnical perspective on urban analytics: The case of city-scale digital twins. *Journal of Urban Technology*, 28(1–2), 263–287. https://doi.org/10.1080/10630732.2020.1798177.

Nordstrom, J. (2016). "A pleasant place for the world to hide": Exploring themes of Utopian play in Ready Player One. *Interdisciplinary Literary Studies*, 18(2), 238–256. https://doi.org/10.5325/intelitestud.18.2.0238.

Poplin, A. (2022). Let's discuss our city! Engaging youth in the co-creation of living environments with digital serious geogames and gamified storytelling. *Environment and Planning B: Urban Analytics and City Science*. https://doi.org/10.1177_23998083221133828.

Roumpani, F. (2022). Procedural cities as active simulators for planning. *Urban Planning*, 7(2), 321–329. https://doi.org/10.17645/up.v7i2.5209.

Therias, A., & Rafiee, A. (2023). City digital twins for urban resilience. *International Journal of Digital Earth*, 16(2), 4164–4190. https://doi.org/10.1080/17538947.2023.2264827.

WEF. (2022, 20 April). Digital twin cities: Framework and global practices. https://www.weforum.org/reports/digital-twin-cities-framework-and-global-practices/.

Wilson, A., & Tewdwr-Jones, M. (2021). *Digital participatory planning: Citizen engagement, democracy, and design* (1st ed.). Routledge. https://doi.org/10.4324/9781003190639.

Young, L. (2021). *Planet city*. URO Publications.

INDEX

Note: *Italic* page numbers refer to figures.

Abercrombie, P. 186
ABM *see* agent-based modelling (ABM)
Abt, C.C. 218
Addis, A. 219, *219*; *The Sumerian Game* 219, *219*
Adelaide Oval *264*
Aerometrix 264–265
agent-based modelling (ABM) 71, 174, 220; AI-Powered Digital Twin for Sarajevo Urban Plan 249, 251–255; geodesign 195–196
AI-Powered Digital Twin for Sarajevo Urban Plan 249, 251–255; data-driven and agent-based simulations 251–253; facilitating decision-making 253–255
Albino, V. 25
Alexander, C. 188
Almere Oosterwold, Netherlands *35*
Alphabet's Sidewalk Labs 14
Altshuller, G. 139
Amsterdam Smart City dashboard 19, 197–198, *197*
Anadol, R. 118
analytic hierarchy processes (AHP) 188
Angelidou, M. 97
Apollo Protocol (2022) 140
application programming interface (APIs) 100
AR *see* augmented reality (AR)
ArcGIS CityEngine 99, *163*, 191
ArcGIS StoryMaps 15, 181
ArcGIS Urban 116
The Architect: Paris 226, *228*
architecture, engineering, and construction (AEC) 99
AR6 Synthesis report 45
Art + Com 53, *53*
artificial intelligence (AI) 94, 175, 263
ASEAN Smart Cities Network 65
The Assessment List on Trustworthy Artificial Intelligence (ALTAI) 113

augmented reality (AR) 32, 78, 80, 124–125; geodesign 199–203; Gothenburg Digital Twin 124
Autodesk University 98

backcasting 67, 91, 178–180; category 179; Transition scenarios 179–180
Bailrigg Garden Village 241
Banathy, B.H. 144
Barcelona Placa Catalunya *34*
Batty, M. 19, 25, 55, 56, *65*, 91, 148, 265; *The Computable City* 33
Beattie, H. 123
Bereitschaft, B. 125, 223
Bergensten, J. 230
Bibri, S.E. 179
big data 94
Big Earth Data 137
Biljecki, F. 60, 62, 76, 99, 114
Bjørvika urban development 120
'black box' testing 221
BladeRunner 2049 192, *192*
Blank, G. 219
Blender 99, 192
Block-by-Block project 31, 160, *161*, 230, 269
Block'hood (Plethora Project) 226, *227*
Bluesky International (2023) 23
Bolivar Water Treatment Plant *149*
Book, K. 274
Borges, J. 183
Braun, J. 220, *224*
British Design Council 144, *144*
British Geological Survey (BGS) 267, *268*
British Ordnance Survey 64
Brown, E. *182*
Bueno, C. 99
Building City Dashboards project *121*
building information modelling (BIM) 60, 91, 97
Building Smart International 23

283

Caldarelli, G. 208
Campagna, M.: Metropolitan City of Cagliari (Italy) case study 20
Campbell, S. 14, 15
CapeReviso project 201
cartography 3, 120; map 54; techniques 26
CASA, City Dashboard 196–197
CAVEs (cave automatic virtual environment) 200
Celestial Software 225
cellular automata (CA) modelling 220–221
Central Area Model, Singapore 273
centralised encompassing system 149
Centre for Digital Built Britain (CDBB) 55, 98, 140
Centre for Spatial Data Infrastructures and Land Administration 48
change model, geodesign 152, 153, 155; agent-based modelling (ABM) 195–196; data dashboards 196–199; land surface models (LSMs) 193–195; procedural stage 190–193
Chevron Corporation 223, 224
China Academy of Information and Communication Technology 54
Chrisman, N. 145
CIMs see city information models (CIMs)
Cisco Systems (2005) 19
Cities: Skylines 125, 218, 233, 238–245, 247–249
Cities: Skylines II 237, 238
Cities XL (Focus Entertainment) 225, 226
Citizen Science 78–79
city-building games 218, 231, 236; Cities XL 225; Micropolis 32, 125, 220–222; SimCity 32, 125, 220–222, 225, 269
city digital twins 97; see also digital twin (DT)
CityEngine 116, 117, 191, 266; ESRI ArcGIS 99, 163, 191; procedural modelling 191
CityGML 22, 23, 60
city information models (CIMs) 91, 96, 98–99, 104, 140, 230, 262, 263
City of Munich 94, 121, 204
City of Zurich Digital Twin 106, 107, 107, 108
City on the Web (2023) 29
City Scope project 195, 196

CITYSTEPS 62, 99
Clark, J.H. 100
climate change 47; adverse impacts 47
Climate Resilience Demonstrator (CReDo) project 115, 116
Cline, E.: Ready Player One film 277, 277
cloud computing 108
CoExist2 109, 109–110
Colding, J. 16
Colossal Order 225, 244
Commuter Flowers, UK 26
Complete Streets Tool 163
computer generated architecture (CGA) 191
computer graphics: geodesign and 145–148
computer vision (CV) techniques 76
contributed geographic information (CGI) 184
CORINE 49
Coulton, P. 217
Council, G. 173
County of London Plan 186, 187
COVID-19 119
Cullen, G. 188
Cureton, P. 102, 142, 182
Currie, A. 68

Daglow, D. 220
Dangermond, J. 141
dashboards 119–122; see also specific dashboards
data: acquisition 21–24; dashboards 196–199; fusion 104; stories 181–182, 206; visualisation 26
Data City Dublin 206–207, 206
data-informed models 25
data ontology 107; CityGML 22–23
Dawkins, O. 198, 206
Debnath, R. 155
decision model, geodesign 152, 154; agent-based modelling (ABM) 195–196; data dashboards 196–199; data stories 181–182; participatory geographic information systems (GIS) 184–187; perception studies 188–190; physical model 204–205
Delaware Basin 18

INDEX

design fiction 67–70, *69*, 218, 263; components 67–68; diegetic prototype 68, *74*, 76; storytelling 68; world-building 68, 71–77
Design with Nature (McHarg) 17, 141
Devisch, O. 32
Digital Beijing (51 World's model) 27, *28*
digital elevation models (DEM) 21
digital games 160, 217, 248, 266; Cities: Skylines 218; Micropolis 32, 125, 220–221, *221*; Minecraft 31, 110, 230, 267, *268*, *270*; SimCity 32, 125, 220–222, 225
Digital geoTwin Vienna 94, *95*
digital replica 4, 58, 95
digital technology 8, 19, 141
digital terrain models (DTM) 21
digital twin (DT) 27; ambition 104; challenges of 55–56; contemporary nature of 98; contextual characteristics 139; environmental 98; geographic information systems and 58–62; human-centric 66; implementation 63; indicative schematic of 100; internal dynamics of 100; pilot studies 110; system of systems 101; technological readiness level for 58, *59*; Uppsala city 116, *117*; urban digital twins 122–128; virtual representation 100–101; visions and attributes of 70; *see also* urban digital twins (UDTs)
Digital Twin Consortium 100, 110
Digital Twin Munich *121*
Digital Twin of Earth 4, 5, *5*, 98
Digital Twins for Smart Cities: Conceptualisation, Challenges, and Practices (Wan, Nochta, Tang, & Schooling) 33
Digital Urban European Twins (DUET) project 110–112, *111*
Division, J. 4; Unknown Pleasures *4*
Donald, B. 145, *146*
Dong, E. 119
Dorffner, L. 95
Double Diamond 144, *144*
DT *see* digital twin (DT)
Dubey, A. 189, *190*
Dublin Dashboard 120, 198
Dunkrel, A. 182

Dunne, A. 67
Dunn, M.: Nottingham City Council 105
Dunn, N. 274
Dutton, G. 3, 5; holographic map of United States *4*
Dyment, D. 219
DynaPlan 120

Eagles' Nests Landscape Park 79
Eikelboom, T. 156, *159*
Eleven Geodesign Method Cards (2024) *175*
EnerPol 251
Engineering Operations game *147*
Enodo Games Inc. *228*
environmental digital twins 98; *see also* digital twin (DT)
Ervin, S.M. 141
ESRI 119, 160; ArcGIS CityEngine 99, 100, *163*, 191; ArcGIS StoryMaps 15, 181; NFrames Photogrammetry software 31
European Space Agency (ESA) 4–5, 98; Digital Twin of Earth 4–5, *5*
EuroSDR Geo-BIM project 32
evaluation model, geodesign: augmented reality (AR) 199–203; land surface models (LSMs) 193–195; procedural stage 190–193
extended reality (XR) 199, 231, 264

Fancher, H. *192*
Feng, S. 246
51 World, Digital Beijing 27, *28*
Fishermans Bend Digital Twin *49*
Fisher, R.A. *52*
Flanders Environment Plan 110
Flanders Regional Mobility Plan 110
Flaxman, M. 141
Flickr 184
forecasting method 55, 58
Forshaw, J. 186
Foster, K. 19, 144, 174
Foundation Data Model (FDM) 55
4D (time)-based holographic mapping 4
Fourth Industrial Revolution 33
Frost, A. 181

285

Fuller, B. 13; World Game *13*, 13–14
Fullerton, T. *271*
Funahashi, T. 62, 99
future city visions 54
futures cones 178
futuring 12–17, 172; in built environment 178; methods in 51

Gallardo, S.F. 181
game engines 32, 110, 199, 236, 254, 266
gameplay 218, 241–244, 247
gamification 30–32, 76; of GIS systems 236; toolkit 125; for urban planning 32
gaming planning systems 229–231
Gamma 195
Geddes, P. 17
Gelernter, D. 27
Gemini Principles 98, 140
generative adversarial neural (GAN) 117
generative artificial intelligence (AI) 150, 265
GeoBIM 99
Geodan, Smart City Amsterdam dashboard 197, *197*
geodesign 17–20, *20*, 48, 51, 141, 262, 266; assessment 151, 174; backcasting 178–180; change model 152, 153, *155*; data stories 181–182; decision model *see* decision model, geodesign; evaluation model 151, 153; framework 109, 142–144, 148–159, 174; impact model *see* impact model, geodesign; intervention 152–153, 174; perception studies 188–190; phases 17, *142*, 144, *144*; platforms 153–154; process model *see* process model, geodesign; representation model *see* representation model, geodesign; scenarios 147–148; stakeholder engagement 112; tools 156, *159*; volunteered geographic information (VGI) 182–184
geographic artificial intelligence (GeoAI) 66
geographic information systems (GIS) 5, 15; cloud-based 60, 77, 93; and digital twins 58–62; in environmental science 137; GIS Science 51; participatory 184–187; real-timeness 103; visualisation of 120
geographic markup language 22

geographic space (geo-scape) 141–142
geo-information: Hamburg's Connected Digital Twins project 140, *140*
geospatial web platforms 114
German Research Foundation Priority Program 183
Getty Institute 117
Gibson, W. 274
GIS *see* geographic information systems (GIS)
Global Power City Index (GPCI) 25
Göteborg 2050 Project 67
Gothenburg Digital Twin 124
Gottwald, S. 184, 185
Greater London Authority 186, *187*
Green, M. *192*
Grieves, M. 91

Haimakainen, H. 244
Hall, P. 6
Halprin, L. 149
Hamburg's Connected Digital Twins project 140, *140*
Haraguchi, M. 62, 99
Harrison, J. 7
Harsia, E. 186
Hartley, E. *102*, *191*
Harvard Graduate School of Design project 117
Harvard Laboratory of Computer Graphics 137, *138*, 145, *146*
Hawken, S. 148, *149*
Haynes, L. 119
Hebei Provincial Planning Institute (HPPI) *157*
He, L. *142*, 156, *157*, 246
Helmer, O. *147*
Hemmersam, P. 120
Heritage Action Zone (HAZ) *182*
Herrenberg digital twin 200
Hessler, J. 3
Hester, R. 156
Hidalgo, C.A. *190*
High Performance Computing Center Stuttgart (HLRS) 200–201, *201*
Hiles, J. *224*
Hines, A. 179, *180*
Hollstein, L.M. 145
Hopkins, D. *221*

INDEX

Houseal Lavigne 117, *118*, 156, *158*, 182, 236, 236, *237*, 265, *265*
Huber, C. 198
Hudson-Smith, A. 123, *124*, 148, 201
Hughes, M. 147
Hunter, D. 125
Huo, Y. 156
Hürzeler, C. 106

IBM (2008) 19
IMD Smart Cities Index 57
IMD-SUTD Smart City Index (SCI) 93
impact assessment 58
impact model, geodesign 152, 153; agent-based modelling (ABM) 195–196; data dashboards 196–199; data stories 181–182; decision 204–205; physical model 204–207; procedural stage 190–193
inequalities 56
information and communication technology (ICT) 19, 57, 93
information design, strategic dashboards 26–27
Information Management Framework (IMF) 55, 98
information systems: adoption 76; geographic *see* geographic information systems (GIS)
inhabitants 80
INSPIRE Directive 45
Institute for Manufacturing 115, *115*
Institute of Public Administration 15, *16*
interactive game 218; *see also specific games*
Intergovernmental Panel on Climate Change (IPCC) 45, 56
International Geodesign Collaboration (IGC) 146–147
International Kriegsspiel Society 217
International Standard Classification of Education (ISCED) 57
Internet of Things (IoT) 14
Ito, K. 76

Jabatan Tanah Dan Survey 22
Jacobs, J. 162, 188
Janssen, R. 156, *159*
JigsAudio 185, *185*
Jing, C. 26

John Hopkins University 119
Joint UK Land Simulator (JULES) 193
Jovanovic, D. 249, *250*

Kahn, H. 146
Kawaii City, Linden Labs *202*
Khemlani, L. 98
Kishino, F. 199, 231
Kitchin, R. 22, 119
Klefbom, S. 78, *81*
Kolbe, T.H. 23
Kontokosta, C. 24, 25
Koven, C.D. *52*
Kumar, B. 63
Kynvin, J. 181

Lady Bug Tools 127
Lake District National Park 247
Lamb, K. 173
Lancaster, UK 238–244, *240*; City Information Model (LCIM) *102*, 266
Land Cover 49; UK Centre for Ecology & Hydrology (UKCEH) 49, *50*
land surface models (LSMs) *52*, 53, 140, 193–195, *194*
Lang, F.: Metropolis film 33, *35*
Latour, B. 217
'layer-cake' method 141
Lee, M.-C. 146
Lego models 273
Lehner, H. 95
Leica City Mapper 22, *23*
Letchworth Garden City (2022) 239
level of detail (LOD) 22–23, 100, 104
Linden Labs *202*
Linde, P. 273
linear progression 54
Liu, X. 113
Living Labs 43; limitations 77–78; maturity model 58–62; urban digital twins and embracing speculation 67–71; world-building 71–77
Living with Machines project 65, *65*
Li, W. 102–103
local planning authorities (LPAs) 229
Löfgren, K. 114

Lovins, A. 178
LSMs *see* land surface models (LSMs)
Lynch, K. 188; *The Perceptual Form of the City* 189

Macdonald, M. 101
machine learning (ML) 64, 251, 265
Malleson, N. 195
Manhattan Island 21; aerial survey 21, *21*
Mankins, J.C. 58
Mapbox API 123
Mapillary 78, *79*, 183
MapReader 64
Markle Foundation 223
Marsal-Llacuna, M. 198
Masoumi, H. 55
Matsuda, K. 274; Hyperreality film 274, *275*
Mattern, S. 198
maturity model 58–62, 99
Maxius Business Simulations 223
May Day 125
Maynooth University 120
McDowell, A. 69
McElvaney, S. 17, 142, 162–164, *163*, 262
McHarg, I.L. 17, *18*, 141; *Design with Nature* 17, 141
McKay, W. 219
Mergardt, M.A. 219
Meridian Water redevelopment project 162
Meskus, M. 67
Metropolis film (Lang) 33, *35*
Micropolis 32, 125, 220, *221*
Mid Cornwall Metro project 269
Milgram, P. 199, 231
Miller, W. 141
Milton, R. *65*, 265
Minecraft 31, 110, 230, 267, *268*, 270
Minority Report 69, *70*
MIT Media Lab 195
M&M project 97
Municipality of Rotterdam, Netherlands 96
MVRDV 33, *35*

Naik, N. *190*
Nass, S. *223*

National Digital Twin (NDT) 55; Climate Resilience Demonstrator (CReDo) 115, *116*
National Mapping Agency data 140
National Planning Policy Framework (NPPF) 229, *229*, 238
National Planning Practice Guidance (NPPG) 229
national sample of cities (NSC) 43–44
National Underground Asset Register (NUAR) 269
nature based solutions (NBS) research 142
Netlogo 195
NetZero production 116
neural radiance field (NeRF) 27, *29*
New London Architecture (NLA) Model 205, *205*
Noardo, F. 32, 99
Nochta, T. 139–140, 263; *Digital Twins for Smart Cities: Conceptualisation, Challenges, and Practices* 33
non-uniform rational B-spline (NURBs) 192
North American Forest Futures Backcasting Scenarios *180*
Nottingham City Council 104, *105*
Nummi, P. 186
NVIDIA 138, *139*

OpenCities application 124
open data 21–24
Open Data Impact for Smart Cities (ODISC) 119
Open Data Institute 119
Open Geospatial Consortium 60
open street map (OSM) 78, 182, 238, 244
Ordnance Survey, UK 100
organisational management 63
Orland, B. 146
OSM *see* open street map (OSM)

Pagani, M. 250, *252*
Page, J. 8
paradigm shift 3, 33, 98, 141, 278
Parikh, D. *190*
Parlikad, A.K. 263
participatory geographic information systems (GIS) 184–187
Participatory Incremental Urban Planning (PIUP) 6
participatory modes 122–123
Pask, G. 30, 31; Kawasaki Project 30–31, *31*

INDEX

Pasmore, W. 68
Pearce, C. 222
perception studies, geodesign 188–190
The Perceptual Form of the City (Lynch) *189*
personal rapid transport (PRT) 116
Persson, M. 230
Phdungsilp, A. 67
Pipers Model Makers *205*
Place Pulse (1.0 and 2.0) 189–190, *190*
Planet Garden 249, *250*
Planning London Datahub 186, 198
planning process 120
Platform Stack Architectural Framework *101*
Plethora Project (Block'hood) 226, *227*
Policy for Integration Architecture 55
polygonal modelling 192
polyphony 186
Poplin, A. 269, *270*
post-disciplinary approaches 7
pre-trained deep learning models 117
Price, C. 31; Fun Palace project 30, *30*; Kawasaki Project 30–31, *31*
process model, geodesign 151, 153; augmented reality (AR) 199–203; backcasting 178–180; land surface models (LSMs) 193–195; participatory geographic information systems 184–187; volunteered geographic information (VGI) 182–184
Projection Augmented Relief Model (PARM) *105*
public participation geographic information system (PPGIS) 16, 106, 184

Quantitative Urban ANalyTics (QUANT) model 64, 65, 193, 265
Queen Elizabeth Olympic Park 123

Raby, F. 67
radio-frequency device (RFID) 186
Raes, L. *111*
Rainforest UK (RFUK) *52*, 53
RAND Corporation 146; Mathematical Analytics Division's (MAD) 217
Rankin, B. 53–54
Raskar, R. *190*
Ready Player One film (Cline) 277, *277*

Reference Data Libraries 55
Regional Innovation 57
regional planning futures 7
ReMap Lima project 193
Ren, C. 29
representation model, geodesign 151, 153, 177; backcasting 178–180; data stories 181–182; participatory geographic information systems (GIS) 184–187; perception studies 188–190; physical model 204–207; volunteered geographic information (VGI) 182–184
Rhino 99
RISE Research Institutes of Sweden 80–82
Robinson, J. 178
Romero, M. *180*
Rosier, J. *159*
Roumpani, F. 193, 266, *267*
Royal Town Planning Institute (RTPI) 231
rule-based change model 174
Ruscha, E. 117

Santa Paravia en Fiumaccio 219
Sarajevo Olympic Center *254*
Sarawak Multimedia Authority 21
Schooling, J.M. 263; *Digital Twins for Smart Cities: Conceptualisation, Challenges, and Practices* 33
Schrotter, G. 106
Schutte, J. *180*
Schwab, K. 33
Scott, G. *159*
Scott, R. *192*
Al-Sehrawy, R. 63
serious games 77, 123, 125, 146, 218; Tirolcraft 269, *270*; *see also specific games*
Shi, W. 113
Short Timeline of Digital Twins *92*
Signorelli, V. 123, *124*
SimCity 32, 125, 220–221, 225
SimCity 2000 game 233
SimHealth 223
SimNimby 222
SimRefinery 223
simulation tools: world-building games and 218–228

289

Singapore Land Authority (SLA) 66, 120, 273
Singleton, A. 113–114
Smart-Cambridge, UK 115
smart cities 24–29; Amsterdam Smart City dashboard 19, 197, 197; Batty's ideas 91; criticisms of 97; development of 27; phases 93–94
Smart Nation Initiative 65
Social Sustainability in Urban Development 80
socio-technical futures 55, 68
Song, K. 29
Souza, L. 99
Spatial Design Competence Center 203
spatial digital twins 98; see also digital twin (DT)
spatial planning 6, 49, 185
speculative design 263, 274, 275
Spielberg, S. 70; Minority Report 69, 70
Spielman, S. 113–114
Srivastava, S.K. 158, 159
Starr, P. 223, 224
Stehle, S. 119
Steinitz, C. 17, 141–144, 152, 155, 174; A Framework for Geodesign: Changing Geography by Design 145
Sterling, D. 67
Stiles, R. 149
storytelling 68
Strategic Foresight Group 180
Sturt, T. 181
The Sumerian Game (Addis) 219, 219
sustainable development goals (SDGs) 43, 56, 239; SDG 11 Explorer 43, 45, 46; urban digital twins (UDTs) 48
Swiss National Museum 79
Sydney dashboard 122
SYMAP (SYnagraphic MAPping) program 145

tabletop games 217; see also specific games
Tallinn Strategy Unit 203
Tang, J.: Digital Twins for Smart Cities: Conceptualisation, Challenges, and Practices 33
Taylor, C. 25; Commuter Flowers, UK 26
technology readiness levels (TRL) 43, 58, 98, 273; analytical/experimental critical function 60; component/breadboard validation 60–61; for digital twins 58, 59; for governance and policy replication 62; levels 60–62; proof-of-concept 60; representative model/prototype system 61; speculative prototypes 60; system prototype in space environment 61; technological maturation 60
Terrestrial Laser Scan 182
Tewdwr Jones, M. 185, 185, 272
The Theory of Inventive Problem Solving 139
Thoreau, H.D. 269
3Dbag 64, 64
three-dimensional (3D) technologies 9; level of detail (LOD) 60; visualisation 9–10
Tikka, E. 67
Tirolcraft 269, 270
Tomlinson, R. 137; Harvard Computer Graphics Week Program 138
Town of Morrisville 117, 118; and Houseal Lavigne 237
Trindade Neves, F. 119
Tufte, E. 120
Tzachor, A. 48, 63, 72

UDTs see urban digital twins (UDTs)
Uhlenkamp, J.-F. 62
UK Coal Industry 68
UN-Habitat Block-by-Block methodology 230
Unique Property Reference Number (UPRN) 100
University of Delaware 16, 16
University of Melbourne 48, 49
Unknown Pleasures (Division) 4
Unreal Engine 99, 117, 236
UN System Framework for Action on Equality 56
Uppsala city 116, 117
urban analytics 24, 112–118
Urban Analytics Lab 183
urban digital twins (UDTs) 11, 27, 31, 43, 48, 50, 91, 137, 261; analytics 112–118; application 145; bottom-up approaches for 43; challenges 62, 262–263, 265–266, 277, 278; component of 266; dashboards 119–122; design fiction 67, 68, 69; digital twin vs. 122–128; embracing speculation 67–71; features 54; flight proven 62; on forecasting methods 55, 58; functioning

INDEX

system 61; fundamental research gap 68–69; geodesign *see* geodesign; human-centric 278; inequalities 56; limitations 77–78; natural synergy in 156; organisational management 63; for policy testing and resource allocation 62; predictive systems 79; socio-technological relationships 97; sustainable development goals (SDGs) 48; system of systems approach 100; systems approach 99; urban analytics 112–118; virtualisation of complexity 98–112; world-building 71–77

urban game continuum 231–238, *232*
urban informatics 24, 113
urban infrastructure 47
urban planning: futures 12–13; gamification for 32; geographic information systems (GIS) 5
Urban Redevelopment Authority (URA) 204, 273, *273*
urban science: futures for 51; global framework of 51
urban settlements 94
Urban Transformation Project Sarajevo (UTPS) 251–255
USDA Forest Service 180
user experience (UX) 125
Utopia: The Creation of a Nation 225, *225*

vapour-worlds 67
Vesselényi, L.S. 78, *81*
VGI *see* volunteered geographic information (VGI)
Villeneuve, D. *192*
Virdee, M. 147
Virtanen, J.-P. 107
Virtual Bradford *103*
Virtual Gothenburg Lab 78, 80–82, 124; CoExist2 *109*, 109–110
Virtual Helsinki 127, 264
virtualisation 104; of complexity 98–112
Virtual London (ViLO) 123, *124*
virtual reality (VR) 124–125
virtual representation 98, 100, 104
visualisation techniques 3; geographic information systems (GIS) 120; 3D model 9–10
volunteered geographic information (VGI) 43, 76, 78, 104, 182–184, 262

Vornhagen, H. 122
Voros, J. 178

Walczak, M. *252, 253*
Walden 269–272, *271*
Waldheim, C. 117
Wan, L. 263; *Digital Twins for Smart Cities: Conceptualisation, Challenges, and Practices* 33
Warner Bros (2017) 192
Warntz, W. 3
Watson, R. 63
Webster, C.W.R. 114
Weiner, M. *147*
Werbach, K. 125
West Cambridge Development Digital Twin 115, *115*
white box testing 221
wicked problems 14, 15, 174
Wilson, A. 185, *185*, 272
Wilson, E.O. 274
Wilson, M.W. 150
Wingtra Drone and Oblique Sony a6100 107
Wong, C. 48–49
World Bank Indicators 57
world-building 68–77; games 218–228
World Economic Forum 54
World Game (Fuller) *13*, 13–14
worldmaking 249
World Settlement Footprint (WSF) 47
Wu, J. 194

Yap, W. 183, *183*
Ye, X. 156
Yigitcanlar, T. 71
Young, L. 274; *Planet City* 274, *276*

Zellner, M. 14, 15
Zeng, X. 29
Zhang, A. 113
Zhang, J. 29
Zhao, Y. 246
Zhou, K. 148, *149*
Zhou, W. 51
Zoan 27, *29*

291

For Product Safety Concerns and Information please contact our EU representative GPSR@taylorandfrancis.com Taylor & Francis Verlag GmbH, Kaufingerstraße 24, 80331 München, Germany

Batch number: 08415860

Printed by Printforce, the Netherlands